ERGEBNISSE DER MATHEMATIK
UND IHRER GRENZGEBIETE

HERAUSGEGEBEN VON DER SCHRIFTLEITUNG
DES
„ZENTRALBLATT FÜR MATHEMATIK"

FÜNFTER BAND
1

SUBHARMONIC FUNCTIONS

BY

TIBOR RADÓ

CHELSEA PUBLISHING COMPANY
231 WEST 29TH STREET, NEW YORK 1, N. Y.

1949

Preface.

A convex function f may be called *sublinear* in the following sense: if a linear function l is $\geq f$ at the boundary points of an interval, then $l \geq f$ in the interior of that interval also. If we replace the terms *interval* and *linear function* by the terms *domain* and *harmonic function*, we obtain a statement which expresses the characteristic property of *subharmonic functions* of two or more variables. This generalization, formulated and developed by F. Riesz, immediately attracted the attention of many mathematicians, both on account of its intrinsic interest and on account of the wide range of its applications. If $f(z)$ is an analytic function of the complex variable $z = x + iy$, then $|f(z)|$ is subharmonic. The potential of a negative mass-distribution is subharmonic. In differential geometry, surfaces of negative curvature and minimal surfaces can be characterized in terms of subharmonic functions. The idea of a subharmonic function leads to significant applications and interpretations in the fields just referred to, and conversely, every one of these fields is an apparently inexhaustible source of new theorems on subharmonic functions, either by analogy or by direct implication. The purpose of this report is *first* to give a detailed account of those facts which seem to constitute the general theory of subharmonic functions, and *second* to present a selected group of facts which seem to be well adapted to illustrate the relationships between subharmonic functions and other theories. Roughly, Chapters I, II, III, V, VI are devoted to the first purpose, while Chapters IV and VII are devoted to the second one. The presentation is formulated for the case of two independent variables, but both the methods and the results remain valid in the general case, except for obvious modifications, unless the contrary is explicitly stated.

Subharmonic functions have a long and interesting history. F. Riesz points out that various methods, due to Poincaré, Perron, Remak in potential theory and to Hartogs and R. Nevanlinna in the theory of functions of a complex variable, are based essentially on the idea of a subharmonic function. The reader should consult Riesz [4], [5] for detailed historical references. Readers interested in the possibilities of applying subharmonic functions may read, for general information, Riesz [4], [5], Beckenbach-Radó [1], [2], Evans [4], Frostman [1].

As it has been observed above, potentials of negative mass-distributions are subharmonic functions, and essentially the converse is also true (see Chapter VI). Thus the theory of subharmonic functions may be interpreted as the study of such potentials based on a few *characteristic properties*, while the methods of potential theory are based on the *representation in terms of definite integrals*. It is very probable that the range of the theory of subharmonic functions, interpreted in this manner, will be considerably extended in the near future. For instance, *the sweeping-out process*, which is fundamental in the recent development of the theory of the capacity of sets (cf. EVANS [4], FROSTMAN [1]) could be easily interpreted in terms of harmonic majorants of subharmonic functions.

Historically, the first generalization of convex functions of a single variable is represented by the *convex functions of several variables*, characterized by the property of being *sublinear* on every straight segment within the domain of definition. While such functions are easily seen to be subharmonic, their theory was developed in connection with problems of an entirely different type. For this reason, the theory of these functions will be included among the topics discussed by W. FENCHEL in a subsequent report of this series.

The reviewer is indebted to G. C. EVANS and S. SAKS for valuable information which he had the privilege to use while preparing this report.

The Ohio State University, March 1937.

TIBOR RADÓ.

Contents.

Chapter I.

Definition and preliminary discussion of subharmonic functions.

1. 1. Let $u(x,y)$ be a function in a domain G (connected open set), such that $-\infty \leq u < +\infty$ in G. That is, $-\infty$ is an admissible value of u, while $+\infty$ is not. Such a function is *subharmonic* in G if it satisfies the following conditions (RIESZ [5], part I, p. 333).

a) u is not identically equal to $-\infty$ in G.

b) u is upper semi-continuous in G. That is, for every point (x_0, y_0) in G and for every number $\lambda > u(x_0, y_0)$ there exists a $\delta = \delta(x_0, y_0, \lambda) > 0$ such that $u(x,y) < \lambda$ for $[(x - x_0)^2 + (y - y_0)^2]^{1/2} < \delta$. Observe that for $u(x_0, y_0) = -\infty$ this condition implies that $u(x,y) \to -\infty$ for $(x,y) \to (x_0, y_0)$.

c) Let G' be any domain comprised in G together with its boundary B'. Let $H(x,y)$ be harmonic in G', continuous in $G' + B'$, and $H \geq u$ on B'. Whenever these assumptions are satisfied, we also have $H \geq u$ in G'.

Superharmonic functions are defined in a similar fashion. A function v is superharmonic in a domain G if the function $u = -v$ is subharmonic there. In the sequel we shall state the results only for subharmonic functions. In n-dimensional Euclidean space subharmonic functions are defined in exactly the same way as in the two-dimensional case. Clearly, a harmonic function is both subharmonic and superharmonic, and conversely.

1. 2. For the sake of accuracy let us observe that F. RIESZ assumed that

a*) $u > -\infty$ on a set everywhere dense in G.

The apparently weaker condition a) in 1. 1 was stated by EVANS [4], part I, p. 230. The following presentation, based partly on unpublished remarks of G. C. EVANS, will show that conditions a), b), c) are equivalent to conditions a*), b), c).

1. 3. Condition b) will be used in the following way. Let S be a closed set comprised in G. Then condition b) implies (HAHN [1]) that there exists a sequence of functions φ_k with the following properties. α) φ_k is continuous on S. β) $\varphi_k \searrow u$ on S, where the symbol $\searrow$ indicates that $\varphi_1 \geq \varphi_2 \geq \cdots$. Conversely, the existence of such a sequence φ_k, for every choice of the closed set S in G, implies that u is upper semi-continuous in G. Take now a domain G' which is comprised in G together with its boundary B'. By what precedes, we have on B' a sequence of continuous functions φ_k such that $\varphi_k \searrow u$ on B'. Suppose that $G' + B'$ is a *Dirichlet region*, that is a region such that for every continuous function f on B' there exists a function H which is harmonic in G', continuous in $G' + B'$, and equal to f on B'. Denote by H_k the solution of the DIRICHLET problem for $G' + B'$ with the boundary condition $H_k = \varphi_k$ on B'. Then $\varphi_1 \geqq \varphi_2 \geqq \cdots$ implies that $H_1 \geqq H_2 \geqq \cdots$ in $G' + B'$. Since $H_k = \varphi_k \geqq u$ on B', it follows from condition c) in 1.1 that $H_k \geqq u$ in $G' + B'$. Summing up: for every Dirichlet region $G' + B'$ in G we have a sequence of functions H_k with the following properties. 1) H_k is continuous in $G' + B'$. 2) H_k is harmonic in G'. 3) $H_k \geqq u$ in $G' + B'$. 4. $H_k \searrow u$ on B'.

According to a fundamental theorem of HARNACK (KELLOGG [1], Chapter X), the property $H_1 \geqq H_2 \geqq \cdots$ implies that in G' the sequence H_k converges either to $-\infty$ everywhere or to a function $\bar{h}$ which is harmonic in G'. In the second case the convergence is uniform on every closed set in G'. The first case can be excluded as soon as $u > -\infty$ at a single point of G'. In the second case, $H_k \geqq u$ in $G' + B'$ implies that $\bar{h} \geqq u$ in G'.

Remark. If u is continuous, we can take $\varphi_k = u$, and $\bar{h}$ is then simply the solution of the Dirichlet problem for $G' + B'$ with the boundary condition $\bar{h} = u$ on B'. In the sequel, the reader interested only in continuous subharmonic functions should always consider this particular choice of φ_k. The reader interested in the general case should glance through the sections 5.1 to 5.4 at this time.

1.4. In the sequel we shall use the following theorems on integration quite frequently. Let there be given, on some range S (curve, domain, etc.) a sequence of functions F_n such that $F_n \searrow F$ on S and $\int_S F_n \geqq A$, where A is a finite constant independent of n (the integrals are taken in the sense of LEBESGUE). Then (see for instance SAKS [5], p. 63 and p. 83) the limit function F is also summable on S and we have $\int_S F = \lim \int_S F_n$.

Consider next a function $f(x, y)$ in a circular disc $D: (x - x_0)^2 + (y - y_0)^2 \leqq r^2$. Introducing polar coordinates we have

$$\iint_D f(x, y)\, dx\, dy = \int_0^r \int_0^{2\pi} f(x_0 + \varrho \cos\varphi, y_0 + \varrho \sin\varphi)\, \varrho\, d\varrho\, d\varphi$$

$$= \int_0^r \left(\int_0^{2\pi} f(x_0 + \varrho \cos\varphi, y_0 + \varrho \sin\varphi)\, d\varphi \right) \varrho\, d\varrho,$$

where we know, by a theorem of TONELLI (SAKS [5], p. 75) that these formulas are certainly valid if $f(x,y)$ is measurable and ≥ 0 in D. More generally, these formulas are valid if f is bounded in one direction, say $f \leq M$ on D, as it follows by applying the preceding remark to the function $M - f$ (quite exactly, whenever one of the three integrals involved exists, the other two exist also and the three integrals are equal to each other).

1. 5. It will be convenient to use the following notations. $C(x_0, y_0; r)$, $D(x_0, y_0; r)$ will refer to the perimeter and to the interior respectively of the circle with centre (x_0, y_0) and radius r, while $R(x_0, y_0; r_1, r_2)$ will refer to the interior of the concentric ring bounded by the circles $C(x_0, y_0; r_1)$ and $C(x_0, y_0; r_2)$. If a function f, defined on $C(x_0, y_0; r)$, is summable as a function of the polar angle φ (where $x = x_0 + r\cos\varphi$, $y = y_0 + r\sin\varphi$), then we shall write

$$L(f; x_0, y_0; r) = \frac{1}{2\pi}\int_0^{2\pi} f(x_0 + r\cos\varphi, y_0 + r\sin\varphi)\, d\varphi.$$

Similarly, if f is defined and summable on $D(x_0, y_0; r)$, we shall write

$$A(f; x_0, y_0; r) = \frac{1}{r^2\pi}\iint_D f(x, y)\, dx\, dy.$$

We have the equivalent formula

$$A(f; x_0, y_0; r) = \frac{1}{r^2\pi}\iint_{\xi^2+\eta^2<r^2} f(x_0 + \xi, y_0 + \eta)\, d\xi\, d\eta.$$

$L(f; x_0, y_0; r)$ and $A(f; x_0, y_0; r)$ are the integral means of f on $C(x_0, y_0; r)$ and $D(x_0, y_0; r)$ respectively.

1. 6. Throughout this Chapter u will denote a function which is subharmonic in a domain G. Suppose that the circle $C(x_0, y_0; r)$ is comprised in G together with its interior and also suppose that $u(x_0, y_0) > -\infty$. Then $L(u; x_0, y_0; r)$ exists and $u(x_0, y_0) \leq L(u; x_0, y_0; r)$ (RIESZ [5], part I, p. 324). To see this, take, as in 1.3, a sequence H_k for the circular disc bounded by $C(x_0, y_0; r)$. We have then $u(x_0, y_0) \leq H_k(x_0, y_0)$ and $H_k(x_0, y_0) = L(H_k; x_0, y_0; r)$ (see KELLOGG [1], p. 82). Hence $L(H_k; x_0, y_0; r) \geq u(x_0, y_0) > -\infty$. By 1.4 it follows that $L(u; x_0, y_0; r)$ exists and that

$$u(x_0, y_0) \leq \lim L(H_k; x_0, y_0; r) = L(u; x_0, y_0; r).$$

1. 7. Under the assumptions of 1. 6 let us consider the disc $D(x_0, y_0; r)$ bounded by $C(x_0, y_0; r)$. We can then apply the result of 1.6 to $C(x_0, y_0; \varrho)$ for $0 < \varrho < r$, and we obtain by 1. 4

$$A(u; x_0, y_0; r) = \frac{2}{r^2}\int_0^r L(u; x_0, y_0; \varrho)\, \varrho\, d\varrho \geq \frac{2u(x_0, y_0)}{r^2}\int_0^r \varrho\, d\varrho = u(x_0, y_0).$$

That is, if $u(x_0, y_0) > -\infty$, then (RIESZ [5], part II, p. 343) u is summable on every disc $D(x_0, y_0; r)$ completely interior to G and $u(x_0, y_0) \leqq A(u; x_0, y_0; r)$.

1. 8. While condition a) in 1.1 states only that $u > -\infty$ for at least one point in G, it follows from 1.7 (according to an unpublished remark of G. C. EVANS) that $u > -\infty$ on a set which is everywhere dense in G. Indeed, if $u(x_0, y_0) > -\infty$, then the summability of u on the disc $D(x_0, y_0; r)$ implies that $u > -\infty$ on this disc with the possible exception of a set of two-dimensional LEBESGUE measure zero. Given then any other point (x', y') in G, we have clearly a finite number of discs $D(x_k, y_k; r_k)$, $k = 0, 1, \ldots, n$, completely interior to G, such that (x_{k+1}, y_{k+1}) is a point of $D(x_k, y_k; r_k)$ for which $u(x_{k+1}, y_{k+1}) > -\infty$, and such that $D(x_n, y_n; r_n)$ contains (x', y'). By 1.7 we have $u > -\infty$ almost everywhere on these discs and hence we have in the vicinity of (x', y') points (x^*, y^*) such that $u(x^*, y^*) > -\infty$.

1. 9. u is summable on every disc $D(x_0, y_0; r)$ completely interior to G (RIESZ [5], part II, p. 343; cf. 1.2). Indeed, by 1.8 we have some disc $D(\bar{x}, \bar{y}; \bar{r})$ completely interior to G, such that $u(\bar{x}, y) > -\infty$ and such that $D(x_0, y_0; r)$ is comprised in $D(\bar{x}, \bar{y}; \bar{r})$. The assertion follows then immediately from 1.7.

1. 10. u is summable on every measurable set S completely interior to G (by completely interior we mean that the limit points of S are also comprised in G). Indeed, by the HEINE-BOREL theorem we can cover the set $S + S'$, where S' is the set of the limit points of S, by a finite number of discs completely interior to G, and the assertion follows then from 1.9 (RIESZ [5], part II, p. 344).

1. 11. u is summable, as a function of the polar angle, on every circle $C(x_0, y_0; r)$ comprised in G together with its interior (RIESZ [5], p. 334). This can be seen by the same reasoning as that used in 1.6, since the assumption $u(x_0, y_0) > -\infty$ was used there only to exclude the possibility $H_k \to -\infty$, and this is excluded now by 1.8.

1. 12. u is summable, as a function of the polar angle, on every circle $C(x_0, y_0; r)$ comprised in G, even if the interior of $C(x_0, y_0; r)$ is not comprised in G (RIESZ [5], part I, p. 338). To see this, take a circle $C(x_0, y_0; r_1)$, $r_1 > r$, such that the ring $R(x_0, y_0; r, r_1)$ is comprised in G together with its boundary. As in 1.3, take a sequence H_k for this ring. By 1.4 the theorem is proved if we show that the sequence $L(H_k; x_0, y_0; r)$ is bounded from below.

Take any smooth JORDAN curve Γ in $R(x_0, y_0; r, r_1)$ which encloses $C(x_0, y_0; r)$. Then the integral

$$\int_{\Gamma} \frac{\partial H_k}{\partial n_e}\, ds,$$

where n_e refers to the outward normal of Γ, is independent of the choice of Γ (KELLOGG [1], p. 212). If we apply this to $C(x_0, y_0; \varrho)$, $r < \varrho < r_1$, then it follows that

$$\varrho \frac{d}{d\varrho} L(H_k; x_0, y_0; \varrho) = a_k$$

and hence

$$L(H_k; x_0, y_0; \varrho) = a_k \log \varrho + b_k$$

where a_k and b_k are constants. On account of the continuity of H_k in the closed ring this formula holds for $r \leq \varrho \leq r_1$. By 1. 8 and 1. 3, H_k converges in the open ring $R(x_0, y_0; r, r_1)$ to a harmonic function $\bar{h}$, the convergence being uniform, in particular, on every circle $C(x_0, y_0; \varrho)$, $r < \varrho < r_1$. Hence $L(H_k; x_0, y_0; \varrho) \to L(\bar{h}; x_0, y_0; \varrho)$ for $r < \varrho < r_1$. That is, the sequence $a_k \log \varrho + b_k$ has a finite limit for every ϱ such that $r < \varrho < r_1$. Clearly, this implies that a_k and b_k converge to finite limits a and b respectively. Then we have $L(H_k; x_0, y_0; r) \to a \log r + b$ and this implies that the sequence $L(H_k; x_0, y_0; r)$ is bounded from below.

1. 13. Using the notations of 1. 12, let us consider $L(u; x_0, y_0; \varrho)$ as a function of ϱ, $r \leq \varrho \leq r_1$. By the theorem of 1. 12, $L(u; x_0, y_0; \varrho)$ actually exists. We observe that by 1. 3 and 1. 4 we have

$$L(u; x_0, y_0; r) = \lim L(H_k; x_0, y_0; r) = a \log r + b,$$
$$L(u; x_0, y_0; r_1) = \lim L(H_k; x_0, y_0; r_1) = a \log r_1 + b.$$

Consider now a third circle $C(x_0, y_0; \varrho)$, $r \leq \varrho \leq r_1$. Then $L(u; x_0, y_0; \varrho) \leq L(H_k; x_0, y_0; \varrho)$ since $u \leq H_k$ in the ring $R(x_0, y_0; r, r_1)$. Hence $L(u; x_0, y_0; \varrho) \leq \lim L(H_k; x_0, y_0; \varrho) = a \log \varrho + b$. As $a \log \varrho + b$ is the (univocally determined) linear function of $\log \varrho$ which is equal to $L(u; x_0, y_0; \varrho)$ for $\varrho = r$ and $\varrho = r_1$, we have the following theorem (RIESZ [5], p. 338).

If the circular ring $0 \leq \varrho_1^2 < (x - x_0)^2 + (y - y_0)^2 < \varrho_2^2$ is comprised in G, then $L(u; x_0, y_0; \varrho)$ is a convex function of $\log \varrho$ for $\varrho_1 < \varrho < \varrho_2$.

1. 14. If we are willing to use somewhat more complicated tools than in the preceding sections, then we can obtain the following more comprehensive result (BRELOT [1], p. 14). If Γ is any sufficiently smooth JORDAN curve in G, then u is summable on Γ as a function of the arc-length (it is not necessary to assume that the interior of Γ is also comprised in G). We modify the proof of BRELOT slightly so as to obtain this theorem directly from the definition of a subharmonic function as given in 1. 1. Let (x_0, y_0) be a point in G such that $u(x_0, y_0) > -\infty$, and assume that Γ does not pass through (x_0, y_0) [clearly, the case of curves passing through (x_0, y_0) can be settled then immediately]. We choose a second smooth JORDAN curve Γ_1 such that the doubly connected domain G' bounded by Γ and Γ_1 is comprised in G together with its boundary, and such that (x_0, y_0) is comprised in G'. As in 1.3, we take a sequence H_k for G'. Since $H_1 \geq H_2 \geq \cdots$ and since all these functions are continuous on $G' + \Gamma + \Gamma_1$, we have a finite constant M such that $u \leq H_k \leq M$, $k = 1, 2, \ldots$, in $G' + \Gamma + \Gamma_1$. If $\mathfrak{G}'(x, y)$ denotes GREEN's function for G' with pole at (x_0, y_0), we have, by applying a general formula (KELLOGG [1], p. 237) to the harmonic function $H_k - M$,

$$H_k(x_0, y_0) - M = \frac{1}{2\pi} \int_\Gamma (H_k - M) \frac{\partial \mathfrak{G}'}{\partial n_i} ds + \frac{1}{2\pi} \int_{\Gamma_1} (H_k - M] \frac{\partial \mathfrak{G}'}{\partial n_i} ds,$$

where n_i refers to the interior normal with respect to G'. We observe that $\partial \mathfrak{G}'/\partial n_i$ has a *positive* minimum $\mu > 0$ on $\Gamma + \Gamma_1$ (cf. BRELOT [1], p. 14). As $H_k - M \leq 0$ on $\Gamma + \Gamma_1$, it follows that

$$H_k(x_0, y_0) - M \leq \frac{\mu}{2\pi} \int_\Gamma (H_k - M) ds.$$

As $H_k(x_0, y_0) \geq u(x_0, y_0)$, it follows finally that

$$\int_\Gamma H_k ds \geq [u(x_0, y_0) - M] \frac{2\pi}{\mu} + M l,$$

where l is the length of Γ. That is, the sequence $\int_\Gamma H_k ds$ is bounded from below. By 1.3 and 1.4 it follows then that u is summable on Γ as function of the arc-length.

1.15. Let G' be a domain comprised in G together with its boundary B'. Suppose that H is harmonic in G', continuous in $G' + B'$ and $H \geq u$ on B'. By condition c) in 1.1 we have then $H \geq u$ in G' also. We shall see now that the sign of equality holds either everywhere or nowhere in G' (RIESZ [5], p. 331). Suppose there is some point (x_0, y_0) in G' such that $u(x_0, y_0) = H(x_0, y_0)$. If r is small, we have then, by 1.7,

$$H(x_0, y_0) = u(x_0, y_0) \leq A(u; x_0, y_0; r) \leq A(H; x_0, y_0; r) = H(x_0, y_0).$$

As $u \leq H$, this clearly implies that $u \equiv H$ in the vicinity of (x_0, y_0). That is, the set of points in G' where $u = H$ is an open set. On account of the upper semi-continuity of u, the set of points in G' where $u < H$ is also open. As the first one of these sets is not empty by assumption, the second one must be empty (since the connected open set G' cannot be the sum of two non-overlapping open sets). Hence $u = H$ everywhere in G'. As u is upper semi-continuous and H is continuous and $\geq u$ in $G' + B'$, it follows immediately that we have $u = H$ on the boundary of G' also.

1.16. As an immediate corollary of the preceding theorem we note the fact that u cannot have a local maximum at a point (x_0, y_0) in G, unless it reduces to a constant in the vicinity of (x_0, y_0), and that u cannot reach its absolute maximum in G unless it reduces to a constant in G.

Chapter II.

Integral means of subharmonic functions.

2. 1. If u is subharmonic in a domain G, then $u(x_0, y_0) \leqq L(u; x_0, y_0; r)$, $u(x_0, y_0) \leqq A(u; x_0, y_0; r)$ for (x_0, y_0) in G and for sufficiently small r (see 1.6, 1.7, 1.9, 1.11). The question arises as to whether these relations are characteristic for subharmonic functions.

2. 2. Denote by K the class of functions u which are defined in a given domain G and satisfy there the following conditions. $\alpha)$ $-\infty \leqq u < +\infty$ and $u \not\equiv -\infty$ in G. $\beta)$ u is upper semi-continuous in G. Denote by K_1, K_2, K_3, K_4 the subclasses of K defined by the following additional requirements. A function u in K belongs to K_1 if for every point (x_0, y_0) in G with $u(x_0, y_0) > -\infty$ we have a $\varrho(x_0, y_0) > 0$ such that for $r < \varrho(x_0, y_0)$ the integral mean $L(u; x_0, y_0; r)$ exists and is $\geqq u(x_0, y_0)$. A function u in K belongs to K_2 if for every point (x_0, y_0) in G with $u(x_0, y_0) > -\infty$ there exists a sequence $r_n \to 0$, depending upon (x_0, y_0), such that $L(u; x_0, y_0; r_n)$ exists and is $\geqq u(x_0, y_0)$, $n = 1, 2, \ldots$. The classes K_3, K_4 are defined in the same way in terms of the integral mean $A(u; x_0, y_0; r)$.

2. 3. On account of 1.6, 1.7, 1.9, 1.11 every function which is subharmonic in G belongs to all four classes K_1, K_2, K_3, K_4. Conversely (LITTLEWOOD [1], p. 189), a function u which belongs to any one of these classes is subharmonic in G. Since conditions a) and b) of 1.1 are satisfied by assumption, we have to verify only the following fact: if G' is a domain comprised in G together with its boundary B', and if H is continuous in $G' + B'$, harmonic in G', and $\geqq u$ on B', then $H \geqq u$ in G' also. If this were not so, then the function $u - H$, which is clearly upper semi-continuous in $G' + B'$, would reach a *positive* maximum M at an *interior* point (x_0, y_0) of $G' + B'$, and the set S of those points (x, y) in $G' + B'$ where $u - H = M$ would be a *closed* set *interior* to $G' + B'$. Since S and B' are closed sets, we have then on S a point (x_1, y_1) whose distance from B' would be a minimum. On every circle $C(x_1, y_1; r)$, with small r, we would have then a whole arc σ_r such that $u - H < M$ on σ_r. Since $u - H \leqq M$ on $C(x_1, y_1; r)$, it follows that

$$L(u; x_1, y_1; r) - H(x_1, y_1) = L(u - H; x_1, y_1; r) < M$$
$$= u(x_1, y_1) - H(x_1, y_1),$$

and hence $L(u; x_1, y_1; r) < u(x_1, y_1)$ for *all* small values of r for which $L(u; x_1, y_1; r)$ exists. A similar reasoning shows that $A(u; x_1, y_1; r) < u(x_1, y_1)$ for *all* small values of r for which $A(u; x_1, y_1; r)$ exists. These conclusions are in obvious contradiction to the assumption that u belongs to one of the classes K_1, K_2, K_3, K_4.

2.4. If u is subharmonic in the domain G, then for fixed (x_0, y_0) the integral mean $L(u; x_0, y_0; r)$ is an increasing function of r as long as the circle $C(x_0, y_0; r)$ is comprised in G together with its interior (RIESZ [5], part I, p. 338). To see this, suppose $C(x_0, y_0; r)$ satisfies the assumption of the theorem and take $r_1 < r$. As in 1.3, take a sequence H_k for the circular disc bounded by $C(x_0, y_0; r)$. We have then $L(H_k; x_0, y_0; r) = H_k(x_0, y_0) = L(H_k; x_0, y_0; r_1) \geqq L(u; x_0, y_0; r_1)$. By 1.3, 1.4 it follows for $k \to \infty$ that $L(u; x_0, y_0; r) \geqq L(u; x_0, y_0; r_1)$.

2.5. We already observed (see 1.13) that as long as the circle $C(x_0, y_0; r)$ remains in a circular ring $R(x_0, y_0; r_1, r_2)$ comprised in G, $L(u; x_0, y_0; r)$ is a convex function of $\log r$ for $r_1 < r < r_2$ and for fixed (x_0, y_0).

2.6. Under the assumptions of 2.5, $L(u; x_0, y_0; r)$ is a continuous function of r. Indeed, if $r_1 < r' < r'' < r_2$, then by reasons of upper semi-continuity u and therefore $L(u; x_0, y_0; r)$ is bounded from above for $r' \leqq r \leqq r''$, and a convex function which is bounded from above is continuous.

2.7. Suppose u is subharmonic in a disc $D(x_0, y_0; \bar{r})$. Then $L(u; x_0, y_0; r) \to u(x_0, y_0)$ for $r \to 0$ (RIESZ [5], part II, p. 344). Indeed, for $r \to 0$ the upper semi-continuity of u implies that $\overline{\lim} L(u; x_0, y_0; r) \leqq u(x_0, y_0)$, while by 1.6 we have $\underline{\lim} L(u; x_0, y_0; r) \geqq u(x_0, y_0)$. Clearly, the reasoning remains valid for $u(x_0, y_0) = -\infty$.

2.8. Similar theorems hold for

$$A(u; x_0, y_0; r) = \frac{2}{r^2} \int\limits_0^r L(u; x_0, y_0; \varrho)\, \varrho\, d\varrho,$$

where it is assumed that the circle $C(x_0, y_0; r)$ is comprised together with its interior in a domain G where u is subharmonic. Since $L(u; x_0, y_0; \varrho)$ is a continuous function of ϱ for $0 \leqq \varrho \leqq r$ (see 2.6, 2.7), we can approximate the above integral by RIEMANN sums, and we obtain the relation (cf. MONTEL [2], p. 49)

$$A(u; x_0, y_0; r) = \lim_{n \to \infty} \sum_{k=1}^{n} \frac{2k}{n^2} L\left(u; x_0, y_0; \frac{k}{n} r\right).$$

As $L(u; x_0, y_0; \varrho)$ is an increasing function of ϱ, it follows immediately that $A(u; x_0, y_0; r) \leqq L(u; x_0, y_0; r)$ (cf. 3.25).

2.9. Under the assumptions of 2.8, $A(u; x_0, y_0; r)$ is an increasing function of r (RIESZ [5], part II, p. 344). This is obvious since the RIEMANN sums used in 2.8 are increasing functions of r by 2.4.

2.10. Under the assumptions of 2.8, we have $A(u; x_0, y_0; r) \to u(x_0, y_0)$ for $r \to 0$ (RIESZ [5], part II, p. 344). The proof is the same as in 2.7.

2.11. Under the assumptions of 2.8, $A(u; x_0, y_0; r)$ is a convex function of $\log r$ (essentially MONTEL [2], p. 49). This is obvious since the RIEMANN sums used in 2.8 are convex functions of $\log r$ by 2.5.

2. 12. We shall say that a function u is of class PL in a domain G if $u \geqq 0$ and if $v = \log u$ is subharmonic there. It is understood that we put $v = -\infty$ for points where $u = 0$. If u is of class PL in G, then u is subharmonic in G (while the converse is obviously false). Indeed, the subharmonic character of $v = \log u$ clearly implies that u satisfies conditions a) and b) of 1.1. Take then any point (x_0, y_0) in G and a small r. As v is subharmonic by assumption, we have $v(x_0, y_0) \leqq L(v; x_0, y_0; r)$ and consequently

$$u(x_0, y_0) = e^{v(x_0, y_0)} \leqq e^{L(v; x_0, y_0; r)} \leqq L(u; x_0, y_0; r).$$

Hence u is subharmonic by 2.3. We used here the inequality

$$\frac{1}{2\pi} \int_0^{2\pi} \log f(\varphi)\, d\varphi \leqq \log \frac{1}{2\pi} \int_0^{2\pi} f(\varphi)\, d\varphi, \qquad f \geqq 0,$$

which is valid whenever the integrals involved have a meaning in the sense of LEBESGUE (for a very elegant proof, see RIESZ [7]).

2. 13. If $u \not\equiv 0$ is $\geqq 0$ and upper semi-continuous in G, then u is of class PL there if and only if $u\, e^h$ is subharmonic for every choice of h in every subdomain G' in which h is harmonic (BECKENBACH [1], for continuous u; the following proof for general u is based on unpublished remarks of S. SAKS). The necessity of the condition being obvious by 2. 12, let us prove that the condition is sufficient. Let G' be any domain comprised in G together with its boundary B'. Let H be continuous in $G' + B'$, harmonic in G', and $v = \log u \leqq H$ on B'. By assumption $u\, e^{-H}$ is subharmonic in G', and $u\, e^{-H}$ is upper semi-continuous even in $G' + B'$, since u is upper semi-continuous and $e^{-H} > 0$ is continuous there. The reasoning of 1. 15 applies therefore to $u\, e^{-H}$ and since $u\, e^{-H} \leqq 1$ on B', we obtain $u\, e^{-H} \leqq 1$ in G' and finally $v = \log u \leqq H$ in G'. That is, $v = \log u$ is subharmonic in G, since v clearly satisfies conditions a) and b) in 1.1 also.

2. 14. If u_1, u_2 are subharmonic in G, then $u_1 + u_2$ is clearly also subharmonic in G, while $u_1 u_2$ will generally not be subharmonic there. On the other hand, the class PL is closed both under addition and multiplication (PRIVALOFF [4]; the following proof is due to S. SAKS). That is, if u_1, u_2 are of class PL in G, then $v = u_1 u_2$ and $w = u_1 + u_2$ are also of class PL. For v this is obvious. As to w, consider any function h which is harmonic in a subdomain G' of G. By 2. 13, $u_1 e^h$ and $u_2 e^h$ are subharmonic in G'. Hence $u_1 e^h + u_2 e^h = w e^h$ is also subharmonic in G'. By 2. 13 it follows that $\log w$ is subharmonic in G.

2. 15. For fixed (x_0, y_0) the function

$$\log r = \log r(x, y; x_0, y_0) = \begin{cases} \log[(x - x_0)^2 + (y - y_0)^2]^{1/2} & \text{for } (x, y) \neq (x_0, y_0), \\ -\infty & \text{for } (x, y) = (x_0, y_0) \end{cases}$$

is a subharmonic function of (x, y) in the whole plane. This follows from the fact that $\log r$ is harmonic for $(x, y) \neq (x_0, y_0)$, while at (x_0, y_0) both the value and the limit of $\log r$ are equal to $-\infty$. If $\alpha > 0$ is a constant, then $\alpha \log r$ is clearly also subharmonic. That is, r^α is of class PL in the whole plane, for $\alpha > 0$. In the case of three independent variables, for instance, we would have

$$-\frac{1}{r} = -\frac{1}{r(x, y, z; x_0, y_0, z_0)}$$

$$= \begin{cases} -\dfrac{1}{[(x - x_0)^2 + (y - y_0)^2 + (z - z_0)^2]^{1/2}} & \text{for } (x, y, z) \neq (x_0, y_0, z_0), \\ -\infty & \text{for } (x, y, z) = (x_0, y_0, z_0) \end{cases}$$

as the simplest unbounded subharmonic function.

2.16. If u is of class PL in a ring $R(x_0, y_0; \varrho_1, \varrho_2)$, then $\log L(u; x_0, y_0; \varrho)$ is a convex function of $\log \varrho$ for $\varrho_1 < \varrho < \varrho_2$ (RIESZ [5], part I, p. 339). To show that a function $f(\varrho)$ has the property that $\log f(\varrho)$ is a convex function of $\log \varrho$, it is sufficient to show that $\varrho^\alpha f(\varrho)$ is a convex function of $\log \varrho$ for every $\alpha > 0$ (see for instance RIESZ [1], p. 6). Let us take any $\alpha > 0$. Put $r = [(x - x_0)^2 + (y - y_0)^2]^{1/2}$. Then, by 2.15 and 2.14, $r^\alpha u$ is of class PL in the ring and hence, by 2.12, $r^\alpha u$ is subharmonic in the ring. By 2.5, $L(u r^\alpha; x_0, y_0; \varrho) = \varrho^\alpha L(u; x_0, y_0; \varrho)$ is therefore a convex function of $\log \varrho$ for $\varrho_1 < \varrho < \varrho_2$.

2.17. If u is of class PL in a disc $D(x_0, y_0; \bar{r})$, then $\log A(u; x_0, y_0; r)$ is a convex function of $\log r$ for $0 < r < \bar{r}$ (essentially MONTEL [2], p. 48). This is obvious since the RIEMANN sums used in 2.8 have (by 2.16) the property that their logarithms are convex functions of $\log r$ for $0 < r < \bar{r}$.

2.18. For various purposes it is important to approximate general subharmonic functions by smooth subharmonic functions. We shall use the following terminology. A function $f(x, y)$ is of class $K^{(0)}$ in a domain G if it is continuous there, and it is of class $K^{(n)}$, $n \geq 1$, if its derivatives of the first n orders are also continuous.

2.19. Let u be subharmonic in a domain G. Consider a domain G' contained in G together with its boundary B'. Put

$$A_r(x, y; u) = A(u; x, y; r) = \frac{1}{r^2 \pi} \iint\limits_{\xi^2 + \eta^2 < r^2} u(x + \xi, y + \eta) \, d\xi \, d\eta$$

(see RIESZ [5], part II, p. 343 and p. 345 for historical references concerning the use of these approximating functions in the theory of harmonic and subharmonic functions). For r fixed and sufficiently small, $A_r(x, y; u)$ is a function of (x, y) which is defined and continuous in G' (cf. 1.10). By 2.1 we have $u(x, y) \leq A_r(x, y; u)$ in G'. As u is bounded from above on every closed set in G, the theorem of TONELLI, referred to in 1.4, may be used to justify the changes in the order of integrations which we are going to carry out. First, from $u(x,y) \leq A_r(x,y;u)$ we obtain by integration $A_{r_1}(x,y;u) \leq A_{r_1}(x,y;A_r) = A_r(x,y;A_{r_1})$. That is (see 2.3) $A_{r_1}(x,y;u)$ is subharmonic. If we put $A_{r_1,r_2}(x,y;u) = A_{r_2}(x,y;A_{r_1})$ and so on, then it follows generally that for $n \geq 1$ the function $A_{r_1,r_2,\ldots r_n}(x,y;u)$ is continuous and subharmonic in G' for small values of $r_1, r_2, \ldots, r_n$ and that $u(x,y) \leq A_{r_1}(x,y;u) \leq A_{r_1,r_2}(x,y;u) \leq \ldots.$ In particular, if we put $A_r^{(n)}(x,y;u) = A_{r,r,\ldots,r}(x,y;u)$, then for small r the function $A_r^{(n)}(x,y;u)$ is continuous and subharmonic in G and we have there $u(x,y) \leq A_r^{(n)}(x,y;u)$.

By 2.8 we have $A_r(x,y;u) \leq A_s(x,y;u)$ for $r \leq s$. By repeated integration we obtain generally $A_r^{(n)}(x,y;u) \leq A_s^{(n)}(x,y;u)$ for $r \leq s$. Finally, by the same reasoning as in 2.7, we obtain $A_r^{(n)}(x,y;u) \to u(x,y)$ for $r \to 0$.

2.20. If $f(x,y)$ is *continuous* in a domain G, then for fixed r the function $A_r(x,y;f) = A(f;x,y;r)$ is easily seen to have continuous derivatives of the first order in the portion of G where it is defined. Indeed, the *four-step rule* for differentiation leads immediately to the formulas

$$\frac{\partial A_r(x,y;f)}{\partial x} = \frac{1}{r\pi} \int_0^{2\pi} f(x + r\cos\varphi,\ y + r\sin\varphi) \cos\varphi\, d\varphi,$$

$$\frac{\partial A_r(x,y;f)}{\partial y} = \frac{1}{r\pi} \int_0^{2\pi} f(x + r\cos\varphi,\ y + r\sin\varphi) \sin\varphi\, d\varphi,$$

which show the continuity of the first derivatives. If f itself has continuous derivatives $f_x = p$, $f_y = q$ of the first order in G, then we have simply

$$\frac{\partial A_r(x,y;f)}{\partial x} = A_r(x,y;p), \qquad \frac{\partial A_r(x,y;f)}{\partial y} = A_r(x,y;q),$$

and the preceding argument shows that $A_r(x,y;f)$ has continuous derivatives of the second order. Generally, if f is of class $K^{(n)}$, then $A_r(x,y;f)$ is of class $K^{(n+1)}$. Applying this to the function $A_r^{(n)}(x,y;u)$ of 2.19, it follows that $A_r^{(n)}(x,y;u)$ is of class $K^{(n-1)}$ in the portion of G in which it is defined.

2.21. For easier reference we sum up the preceding remarks in the following *approximation theorem*. If u is subharmonic in a domain G, then the sequence $u_k^{(3)}(x,y) = A_{1/k}^{(3)}(x,y;u)$, $k = 1, 2, \ldots$, has the following properties. Let G' be any domain comprised in G together with its boundary. Then for large k the function $u_k^{(3)}$ is defined, subharmonic and of class $K^{(2)}$ in G' and $u_k^{(3)} \searrow u$ in G'.

Actually, the integral means $A_r^{(n)}(x,y;u)$ are smoother than it appears from the preceding statements. More precise information could be obtained easily from 6.22.

2.22. For continuous u it follows from 2.19 and 2.20 that for large k the function $u_k^{(2)}(x,y) = A_{1/k}^{(2)}(x,y;u)$ is already of class $K^{(2)}$ in G' and that $u_k^{(2)} \to u$ *uniformly* in G'.

2.23. Suppose that the function u of 2.21 happens to be *harmonic* in a domain G' comprised in G together with its boundary B'. Take any closed set S in G' and denote by δ the shortest distance of S and of B'. Then for $r < 1/(3\,\delta)$ we have, by the mean-value property of harmonic functions, $u_k^{(3)} = u$ on S and hence $\Delta u_k^{(3)} = 0$ on S, $\Delta = \partial^2/\partial x^2 + \partial^2/\partial y^2$.

2.24. Denote by S a closed bounded set in the domain G in which u is subharmonic. Then the functions $u_k^{(3)}$ of 2.21 satisfy for large k an inequality

$$0 \leqq \iint\limits_S \Delta u_k^{(3)}(x,y)\,dx\,dy < M,$$

where M is a finite constant (RIESZ [**5**], part II, p. 353). To see this, observe that S can be covered by a finite number of closed circular discs (HEINE-BOREL theorem) and that therefore it is sufficient to consider the case when S is a closed circular disc D, with radius r and centre (x_0, y_0), comprised in G. We have then by GREEN's identity

$$\iint\limits_D \Delta u_k^{(3)}(x,y)\,dx\,dy = \int\limits_{C_r} \frac{\partial u_k^{(3)}}{\partial u_e}\,ds = \frac{dL\,(u_k^{(3)};\,x_0,\,y_0;\,r)}{d\log r}, \quad C_r = C\,(x_0,\,y_0;\,r).$$

Denote by I_k the common value of these expressions. Take r_1 slightly larger than r. Write $L_k(r)$ for $L(u_k^{(3)};x_0,y_0;r)$ and $L(r)$ for $L(u;x_0,y_0;r)$. Since $L_k(r)$ is a convex function of $\log r$ by 2.5 and 2.21, we have

$$I_k \leqq \frac{L_k(r_1) - L_k(r)}{\log r_1 - \log r}.$$

By 2.21, 1.4, 2.4 it follows for $k \to \infty$ that

$$0 \leqq \overline{\lim}\, I_k \leqq \frac{L(r_1) - L(r)}{\log r_1 - \log r} < +\infty$$

and the theorem is proved.

Chapter III.

Criteria and constructions for subharmonic functions.

3.1. If u is of class $K^{(2)}$ (see 2.18) in a domain G and if $\Delta u = \partial^2 u/\partial x^2 + \partial^2 u/\partial y^2 > 0$, then u is subharmonic. Similarly, if $\Delta u < 0$, then u is superharmonic. Suppose, for instance, that $\Delta u > 0$ in G. Let G' be a domain comprised in G together with its boundary B'. Suppose that H is continuous in $G' + B'$, harmonic in G', and $\geq u$ on B'. We have to show that $H \geq u$ in G' also. Suppose this is not true. Then $v = u - H$ reaches its maximum at an interior point (x_0, y_0), and we should have there $\Delta v \leq 0$, while by assumption $\Delta v = \Delta u - \Delta H = \Delta u > 0$ in G.

3. 2. If u is of class $K^{(2)}$ in G, then u is subharmonic in G if and only if $\Delta u \geq 0$ there (RIESZ [5], part I, p. 335). Proof. Suppose first that u is subharmonic in G. If $\Delta u < 0$ at some point (x_0, y_0) of G, then u is superharmonic in a vicinity G' of (x_0, y_0), on account of 3. 1. Then u is both subharmonic and superharmonic in G', and hence (see 1. 1) u is harmonic there. This is impossible, since $\Delta u < 0$ in the vicinity of (x_0, y_0). Suppose second that $\Delta u \geq 0$ in G. If ε is a positive constant, then the function $u^* = u + \varepsilon (x^2 + y^2)$ satisfies the condition $\Delta u^* > 0$ in G. Hence u^* is subharmonic in G by 3. 1. For $\varepsilon \to 0$ the function u^* converges to u uniformly in G, and the subharmonic character of u is then a consequence of the following theorem.

3. 3. Let u be defined in a domain G. Suppose that there exists a sequence u_k with the following properties. If G' is any domain comprised in G together with its boundary, then for large k the function u_k is defined and subharmonic in G' and $u_k \to u$ uniformly in G'. Then u is subharmonic in G. Briefly, the uniform limit of subharmonic functions is subharmonic (RIESZ [5], part I, p. 335). Observe first that the assumptions clearly imply the upper semi-continuity of u. If (x_0, y_0) is any point in G, then we have, by 2. 1, for large k and small r, $u_k(x_0, y_0) \leq L(u_k; x_0, y_0; r)$ and for $k \to \infty$ it follows that $u(x_0, y_0) \leq L(u; x_0, y_0; r)$. The subharmonic character of u follows then from 2. 3.

3. 4. If a, b are two real numbers, including $\pm \infty$, then $\overline{a, b}$ will denote the larger one of a, b if $a \neq b$ and the common value of a, b if $a = b$. We have then the theorem: if u_1, u_2 are subharmonic in G, then their upper envelope $u = \overline{u_1, u_2}$ is also subharmonic in G (RIESZ [5], part I, p. 335). Proof. u clearly satisfies conditions a) and b) in 1. 1. Let $C(x_0, y_0; r)$ be any circle comprised in G together with its interior. As u_1 and u_2 are upper semi-continuous, we have a finite constant M such that $u_1 < M$, $u_2 < M$ on $C(x_0, y_0; r)$. We have then also $u < M$ on $C(x_0, y_0; r)$, while by definition $u \geq u_1$. Thus u is comprised between two functions, namely u_1 and M, which are summable on $C(x_0, y_0; r)$ as functions of the polar angle. Hence u is also summable on $C(x_0, y_0; r)$ as a function of the polar angle. At (x_0, y_0) we have either $u = u_1$ or $u = u_2$. If $u(x_0, y_0) = u_1(x_0, y_0)$, for instance, then $u(x_0, y_0) = u_1(x_0, y_0) \leq L(u_1; x_0, y_0; r) \leq L(u; x_0, y_0; r)$. Hence u is subharmonic by 2. 3. A similar reasoning shows that the upper envelope of any finite number of subharmonic functions is subharmonic.

3.5. Let F denote a family of infinitely many functions which are subharmonic in a domain G. Suppose that F is a normal family (that is, every infinite sequence of functions of F contains a uniformly convergent subsequence). If $\bar{u}(x,y)$ denotes the largest cluster-value of the values at (x,y) of all the functions of F, then $\bar{u}(x,y)$ is subharmonic in G (Montel [2], p. 38, for continuous functions; Malchair [1], p. 11, for general subharmonic functions). The proof is similar to that in 3.4.

3.6. Let there be given, in a domain G, a sequence u_n with the following properties. If G' is any domain comprised in G together with its boundary, then for large n the functions $u_n, u_{n+1}, \ldots$ are defined and subharmonic in G' and $u_n \geq u_{n+1} \geq \cdots$ in G'. Then either $u_n \to -\infty$ in G or u_n converges in G to a subharmonic function (Riesz [5], part I, p. 335; cf. 1.2). Proof. Put $\lim u_n = u$ and suppose that $u \not\equiv -\infty$ in G. Clearly, $-\infty \leq u < +\infty$ and u is upper semi-continuous in G. Let (x_0, y_0) be any point in G such that $u(x_0, y_0) > -\infty$. We have then, for large n and small r, $-\infty < u(x_0, y_0) \leq u_n(x_0, y_0) \leq L(u_n; x_0, y_0; r)$ and this implies, by 1.4, that $L(u; x_0, y_0; r)$ exists and is $\geq u(x_0, y_0)$. Hence u is subharmonic by 2.3.

3.7. Suppose that u is upper semi-continuous and $-\infty \leq u < +\infty$ in a domain G and that for every (x_0, y_0) in G the integral mean $L(u; x_0, y_0; r)$ exists for small r. If

$$\varlimsup_{r \to 0} \frac{1}{r^2} \left(L(u; x_0, y_0; r) - u(x_0, y_0) \right) \geq 0$$

for every point (x_0, y_0) in G, then u is subharmonic in G (Saks [3], p. 190; this is a generalization of a theorem of Blaschke [1] on harmonic functions). Proof. Observe that the function $u_n = u + x^2/n$, n a positive integer, belongs to the class K_2 defined in 2.2 and apply 2.3 and 3.6. A similar reasoning shows that if

$$\varlimsup_{r \to 0} \frac{1}{r^2} \left(A(u; x_0, y_0; r) - u(x_0, y_0) \right) \geq 0$$

for every point (x_0, y_0) in G, then u is subharmonic in G (Szpilrajn [1], p. 589). For further theorems of this type see the remarks of Saks ([4], p. 382), and see also Kozakiewicz ([1], pp. 5—6).

3.8. Combining 2.21, 3.6 and 3.2 we see that the class of subharmonic functions consists *first* of all functions with continuous second derivatives which satisfy the condition $\Delta = \partial^2/\partial x^2 + \partial^2/\partial y^2 \geq 0$, and *second* of the limits of decreasing sequences of such functions (limits $\equiv -\infty$ being excluded).

3.9. Similarly, by 2.22, the class of *continuous* subharmonic functions consists *first* of the functions with $\Delta \geq 0$ as before, and *second* of the *uniform* limits of such functions.

3. 10. If $u_1, u_2, \ldots, u_n$ are subharmonic in G and if $\alpha_1, \alpha_2, \ldots, \alpha_n$ are non-negative constants, then obviously $\alpha_1 u_1 + \alpha_2 u_2 + \cdots + \alpha_n u_n$ is also subharmonic in G (RIESZ [5], part I, p. 335).

3. 11. The next few theorems will be concerned with relations between subharmonic functions and convex functions. These theorems have the common feature that they can be proved by very simple computations if the functions involved are sufficiently smooth. It is then natural to treat the general case by approximation in terms of integral means (see 2. 21). As a matter of fact, in the theory of subharmonic functions this method of approximation was first used in connection with a problem of this type (the theorem discussed in 3. 12).

3. 12. Let $u \geqq 0$ be upper semi-continuous and $< + \infty$ in a domain G. Then $v = \log u$ is subharmonic in G if and only if $e^{\alpha x + \beta y} u$ is subharmonic there for every choice of the constants α, β (MONTEL [2], p. 39 for smooth u; RADÓ [2], for continuous u), Proof. The necessity of the condition follows immediately from 2. 12. To prove the sufficiency, suppose that $e^{\alpha x + \beta y} u = w$ is subharmonic for every choice of the constants α, β and call this property *the property* (M). If u is positive and of class $K^{(2)}$ in G, then we have $\varDelta w \geqq 0$. Explicitly:
$$\varDelta w = e^{\alpha x + \beta y} [\varDelta u + 2\alpha u_x + 2\beta u_y + (\alpha^2 + \beta^2) u] \geqq 0$$
for every choice of α, β. As the quantity in the bracket is a quadratic function of α, β, we obtain readily the inequality $u \varDelta u - (u_x^2 + u_y^2) \geqq 0$, which shows that $\log u$ is subharmonic. Indeed, we have $\varDelta \log u = [u \varDelta u - (u_x^2 + u_y^2)]/u^2$. If u, still of class $K^{(2)}$, satisfies only the condition $u \geqq 0$, consider first $u_n = u + 1/n$, n a positive integer, and apply 3. 3. Suppose now that u has only the properties specified in the statement of the theorem. It follows then immediately that the function $A_r^{(3)}(x, y; u)$ of 2. 19 also possesses the property (M). As $A_r^{(3)}(x, y; u)$ is of class $K^{(2)}$, $\log A_k^{(3)}(x, y; u)$ is subharmonic by what precedes, and as $\log A_r^{(3)}(x, y; u) \searrow \log u$ for $r \searrow 0$, the subharmonic character of $\log u$ follows by 3. 6.

3. 13. Suppose that $f(t)$ is convex and increasing (and therefore continuous) for $t_1 < t < t_2$ and that $u(x, y)$ is subharmonic in a domain G. If $t_1 < u < t_2$ in G, then $v = f(u)$ is subharmonic in G (MONTEL [2], p. 42, for smooth functions; BRELOT [1], p. 16, for the general case). Proof. If f and u are smooth, we have $\varDelta v = f''(u)(u_x^2 + u_y^2) + f'(u) \varDelta u \geqq 0$, since $f' \geqq 0$, $f'' \geqq 0$, $\varDelta u \geqq 0$ by assumption. The general case can be treated either by approximation (BRELOT, l. c.) or also directly as follows. Let (x_0, y_0) be any point in G. Since u is subharmonic, we have $u(x_0, y_0) \leqq L(u; x_0, y_0; r)$ for small r. Since f is increasing, it follows that $v(x_0, y_0) = f(u(x_0, y_0)) \leqq f(L(u; x_0, y_0; r))$. Since f is convex, we have $f(L(u; x_0, y_0; r)) \leqq L(f(u); x_0, y_0; r) = L(v; x_0, y_0; r)$ (see for instance PÓLYA-SZEGŐ [1], p. 52, problem 71, where the inequality is stated in a somewhat less general form). Thus $v(x_0, y_0) \leq L(v; x_0, y_0; r)$ and hence v is subharmonic by 2.3 (the further assumptions stated there being obviously satisfied in the present case).

3.14. Suppose that $f(t)$ is convex and continuous for $t_1 < t < t_2$ and that $h(x, y)$ is harmonic in a domain G. If $t_1 < h < t_2$ in G, then $v = f(h)$ is subharmonic in G (see 3.13 for references). The proof is the same as in 3.13.

3.15. The theorem of 3.12 can be restated as follows. Suppose that v is upper semi-continuous in a domain G and satisfies there the conditions $-\infty \leq v < +\infty$, $v \not\equiv -\infty$. If $e^{\alpha x + \beta y + v}$ is subharmonic for every choice of the constants α, β, then v is subharmonic, and the converse is also true. KIERST (see SAKS [3], p. 187) raised the following question. For what functions $f(t)$ is it true that whenever $f(\alpha x + \beta y + v)$ is subharmonic for every choice of the constants α, β, then it follows that v is subharmonic. According to 3.12, $f(t) = e^t$ is such a function. KIERST found the following curious theorem. If $f(t)$ for $-\infty < t < +\infty$ and $v(x, y)$ for (x, y) in a domain G have continuous derivatives of the second order, if $f'(t) > 0$, and if $f(\alpha x + \beta y + v)$ is subharmonic in G for every choice of the constants α, β, then v is subharmonic in G. That is, if we restrict ourselves to smooth functions, then every strictly increasing function $f(t)$ has the desired property. Proof. Put $w = f(\alpha x + \beta y + v)$. The assumption that w is subharmonic is expressed by the inequality

$$\Delta w = f''(\alpha x + \beta y + v)[(v_x + \alpha)^2 + (v_y + \beta)^2] + f'(\alpha x + \beta y + v)\Delta v \geq 0.$$

Consider any point (x_0, y_0) in G and choose $\alpha = -v_x(x_0, y_0)$, $\beta = -v_y(x_0, y_0)$. As $f' > 0$, it follows that $\Delta v \geq 0$ at (x_0, y_0). Hence v is subharmonic by 3.2. It is not known at present whether the assumptions concerning the smoothness of f and v are necessary for the validity of the theorem.

3.16. SAKS [3] observed that the functions $f(t)$ for which the theorem of 3.15 is non-vacuous are less general than it would appear from the statement of that theorem. Suppose that 1) $f(t)$ is continuous with its first and second derivatives for $-\infty < t < +\infty$, 2) $f'(t) > 0$ and 3) there exists, in some domain G, a function $v(x, y)$ of class $K^{(2)}$ such that $f(\alpha x + \beta y + v)$ is subharmonic for every choice of the constants α, β. Then $f(t)$ is convex. As in 3.15, the proof follows by a simple discussion of the explicit expression of $\Delta f(\alpha x + \beta y + v)$. Again, it is not known whether the theorem remains true without the restrictions concerning the smoothness of f and v. SAKS (l. c.) modified the problem by introducing a third parameter γ, and studied the situation where $f(\alpha x + \beta y + \gamma + v)$ is subharmonic for every choice of the constants α, β, γ. He found that as a consequence of the increased number of the parameters the assumptions concerning the derivatives of f and v can be dropped. He obtained the following theorem. Suppose that 1) $f(t)$ is continuous for $-\infty < t < +\infty$, 2) $v(x,y)$ is continuous in a domain G, and 3) $f(\alpha x + \beta y + \gamma + v)$ is subharmonic in G for every choice of the constants α, β, γ. Then $f(t)$ is necessarily convex and furthermore one of the following four statements is true. I. $f(t)$ is constant. II. $v(x,y)$ is harmonic. III. $v(x,y)$ is subharmonic in G and $f(t)$ is increasing for $-\infty < f < +\infty$ [that is, $f(t_1) \le f(t_2)$ for $t_1 < t_2$]. IV. $v(x,y)$ is superharmonic in G and $f(t)$ is decreasing for $-\infty < t < +\infty$. If f and v are sufficiently smooth, then this theorem can be proved by a simple argument similar to that used in 3. 15. The hope that the general case can be treated by approximations has not materialized so far. At any rate, the proof given by SAKS is essentially a direct proof. The ingenious details of the proof cannot be reproduced here.

3. 17. Suppose that 1) $f(t)$ is continuous for $t_1 < t < t_2$, 2) $h(x,y)$ is harmonic in a domain G, 3) $t_1 < h < t_2$ in G, and 4) $u = f(h)$ is subharmonic in G. Then $f(t)$ is convex in the interval $m < t < M$, where m and M denote the greatest lower bound and the least upper bound of h in G (MONTEL [2], p. 43, under certain restrictions; SAKS [2], for the general case). Proof. It is clearly assumed that h is not constant. If the theorem is false, then there exists a linear function $at + b$ such that $g(t) = f(t) + at + b$ reaches a proper local maximum at a certain point t_0, $m < t_0 < M$. That is, $g(t) \le g(t_0)$ and $g(t)$ is not constant in the vicinity of t_0. The function h takes on the value t_0 at some point (x_0, y_0) in G, since $m < t_0 < M$, and h takes on all values t close to t_0 in the vicinity of (x_0, y_0), since h is not constant. The function $u = g(h) = f(h) + ah + b$ has then a local maximum at (x_0, y_0) without reducing to a constant in the vicinity of (x_0, y_0). On account of 1. 16 this is however impossible since u is clearly subharmonic.

3. 18. We proceed to quote a few applications of the preceding theorems. Let u be subharmonic in a circular disc $D(x_0, y_0; \varrho)$. Put $[(x - x_0)^2 + (y - y_0)^2]^{1/2} = r$ and consider the function $l(x,y) = L(u; x_0, y_0; r)$. Then $l(x,y)$ is constant on every circle $C(x_0, y_0; r)$, $0 < r < \varrho$, and we can write $l(x,y) = \lambda(\log r)$. By 2. 6, $\lambda(\log r)$ is a continuous, increasing and convex function of $\log r$ for $0 < r < \varrho$. Hence, by 3. 13, $l(x,y)$ is subharmonic in $D(x_0, y_0; \varrho)$, except possibly at (x_0, y_0). To discuss the point (x_0, y_0), observe that $l(x,y)$ is continuous there by 2. 6 and 2. 7, and that $L(l; x_0, y_0; \sigma) = L(u; x_0, y_0; \sigma) \ge u(x_0, y_0) = l(x_0, y_0)$ since u is subharmonic. Hence $l(x,y)$ is subharmonic in $D(x_0, y_0; \varrho)$ (essentially MONTEL [2], p. 48).

3. 19. Under the assumptions of 3. 18 the function $a(x,y) = A(u; x_0, y_0; r)$ is also subharmonic in $D(x_0, y_0; \varrho)$ (see reference in 3. 18). The proof is the same as in 3. 18.

3.20. Under the assumptions of 3.18 the function u will reach a maximum $M(r)$ on every circle $C(x_0, y_0; r)$, $0 < r < \varrho$. Put $M(0) = u(x_0, y_0)$ and consider the function $\mu(x, y) = M(r)$, $r = [(x - x_0)^2 + (y - y_0)^2]^{1/2}$. Then $\mu(x, y)$ is subharmonic in $D(x_0, y_0; \varrho)$ (essentially MONTEL [2], p. 41). On account of 3.14 this theorem is, except for minor details, a consequence of the fact that $M(r)$ is a convex function of $\log r$ (essentially RIESZ [1]). To prove this property of $M(r)$ take r_1, r_2 such that $0 < r_1 < r_2 < \varrho$. We have to show that $M(r) \leqq a \log r + b$ for $r_1 < r < r_2$, where $a \log r + b$ is the linear function of $\log r$ which takes on the values $M(r_1)$, $M(r_2)$ for $r = r_1$, $r = r_2$ respectively. Consider the function $H(x, y) = a \log[(x - x_0)^2 + (y - y_0)^2]^{1/2} + b$. Then H is harmonic and we have $H \geqq u$ on the boundary of the circular ring $R(x_0, y_0; r_1 \ r_2)$ by the definition of $M(r)$ and of a, b. Hence (see 1.1, condition c)) we have also $H \geqq u$ on $C(x_0, y_0; r)$ for $r_1 < r < r_2$ and consequently $M(r) \leqq a \log r + b$ for $r_1 < r < r_2$.

3.21. If in the statements of the theorems in 3.18 to 3.20 we assume that u is of class PL (see 2.12), then it follows immediately that the functions $l(x, y)$, $a(x, y)$, $\mu(x, y)$ are also of class PL (see 3.18 to 3.20 for references).

3.22. Suppose that 1) $u \not\equiv -\infty$, 2) $-\infty \leqq u < +\infty$ and 3) u is upper semi-continuous in a domain G. If u is a convex function of x for every fixed value of y and a convex function of y for every fixed value of x, then u is subharmonic in G (MONTEL [2], p. 37, for continuous u; MALCHAIR [1], p. 7, for the general case). Proof. Take any circle $C(x_0, y_0; r)$ comprised in G together with its interior. Take a sequence of functions g_ν such that g_ν is continuous and $g_\nu \searrow u$ for $(x - x_0)^2 + (y - y_0)^2 = r^2$ (cf. 1.3). The convexity properties of u and the inequality $u \leqq g_\nu$ imply that

$$u(x_0, y_0) \leqq \tfrac{1}{4}[g_\nu(x_0 + h, y_0 + k) + g_\nu(x_0 - h, y_0 + k) + g_\nu(x_0 - h, y_0 - k)$$
$$+ g_\nu(x_0 + h, y_0 - k)]$$

for $h^2 + k^2 = r^2$. Subdivide $C(x_0, y_0; r)$ into n equal arcs by points $(x_0 + h_i, y_0 + k_i)$, $i = 1, \ldots, n$. Write the preceding inequality for $h_i, k_i, i = 1, \ldots, n$. After addition, it follows for $n \to \infty$ that $u(x_0, y_0) \leqq L(g_\nu; x_0, y_0; r)$. By 1.4 this implies that $L(u; x_0, y_0; r)$ exists and is $\geqq u(x_0, y_0)$. Hence u is subharmonic by 2.3.

As a special case of the preceding theorem, a function $u(x, y)$ is subharmonic if the surface $z = u(x, y)$ is convex in the sense that it is intersected in a convex curve by every plane which is parallel to the z-axis. Functions u with this property were the first ones to be considered as generalizations of convex functions of a single variable. Subharmonic functions with such special convexity properties were studied in detail by MONTEL [3] and by VALIRON [1].

3. 23. If $u \geq 0$ is subharmonic in G, then for $\alpha > 1$ the function $v = u^\alpha$ is also subharmonic, since t^α is increasing and convex if $\alpha > 1$ (cf. 3. 13). More generally, if $u_1, \ldots, u_n$ are subharmonic and ≥ 0 in G and if $\alpha > 1$, then $u = (u_1^\alpha + \cdots + u_n^\alpha)^{1/\alpha}$ is also subharmonic. If $u_1, \ldots, u_n$ are of class $K^{(2)}$, then the relation $\Delta u \geq 0$ reduces, by direct computation, to an inequality which is easily recognized as an immediate consequence of the inequality of SCHWARZ. In the general case the theorem follows then immediately by approximating $u_1, \ldots, u_n$ by integral means (see 2. 21).

3. 24. If u is of class PL in G (see 2. 12), then it follows immediately from 2. 12 that u^α is subharmonic for $\alpha > 0$. Conversely, if $u \geq 0$ and $u \not\equiv 0$ in G and if u^α is subharmonic for every $\alpha > 0$, then u is of class PL (cf. MONTEL [2], p. 24). Indeed, $v_n = n(u^{1/n} - 1)$ is then clearly subharmonic for $n = 1, 2, \ldots$, and $v_n \searrow \log u$. Hence $\log u$ is subharmonic by 3. 6.

3. 25. We observed (see 2. 8) that if u is subharmonic in G, then we have $A(u; x_0, y_0; r) \leq L(u; x_0, y_0; r)$ whenever the circle $C(x_0, y_0; r)$ is comprised in G together with its interior. We shall consider now two theorems concerning the characterization of subharmonic functions in terms of inequalities involving only integral means. To simplify the statements we restrict ourselves to the case of continuous functions. We have then the theorem: if u is continuous in G, then u is subharmonic there if and only if $A(u; x_0, y_0; r) \leq L(u; x_0, y_0; r)$ whenever the circle $C(x_0, y_0; r)$ is comprised in G together with its interior (BECKENBACH and RADÓ [2], p. 668). By 2. 8 the condition is necessary. To prove its sufficiency, assume first that u is of class $K^{(2)}$. Let (x_0, y_0) be any point in G. The TAYLOR expansion yields then $L(u; x_0, y_0; \varrho) = u_0 + \frac{1}{4} \varrho^2 (r_0 + t_0) + \sigma_1$, $\quad A(u; x_0, y_0; \varrho) = u_0 + \frac{1}{8} \varrho^2 (r_0 + t_0) + \sigma_2$, where u_0, r_0, t_0 are the values of u, u_{xx}, u_{yy} at (x_0, y_0), and $\sigma_i / \varrho^2 \to 0$ for $\varrho \to 0$, $i = 1, 2$. The inequality $A(u; x_0, y_0; \varrho) \leq L(u; x_0, y_0; \varrho)$ implies that $r_0 + t_0 \geq 8(\sigma_2 - \sigma_1)/\varrho^2$, and for $\varrho \to 0$ it follows that $r_0 + t_0 \geq 0$. Hence u is subharmonic by 3. 2. If u is only continuous, then the theorem follows by approximating u by integral means (see 2. 21).

3. 26. Similarly, if u is continuous and positive in G, then $\log u$ is subharmonic there if and only if $[A(u^2; x_0, y_0; \varrho)]^{1/2} \leq L(u; x_0, y_0; \varrho)$ whenever the circle $C(x_0, y_0; \varrho)$ is comprised in G together with its interior (BECKENBACH and RADÓ [2], p. 665). The sufficiency of the condition is proved by the method used in 3. 25. To prove the necessity, suppose that $\log u$ is subharmonic in G. Take any circle $C(x_0, y_0; \varrho)$ comprised in G together with its interior. Let h be the harmonic function in $C(x_0, y_0; \varrho)$ which coincides with $\log u$ on $C(x_0, y_0; \varrho)$. Denote by g the conjugate harmonic function, and put $f(z) = e^{h+ig}$, $z = x + iy$. According to a theorem of CARLEMAN [1], we have then $[A(|f|^2; x_0, y_0; \varrho)]^{1/2} \leq L(|f|; x_0 \, y_0; \varrho)$. Since $|f| = e^h$, we can write this inequality in the form $[A(e^{2h}; x_0, y_0; \varrho)]^{1/2} \leqq L(e^h; x_0, y_0; \varrho)$. As $\log u$ is subharmonic, it follows that

$$[A(u^2; x_0, y_0; \varrho)]^{1/2} \leqq [A(e^{2h}; x_0, y_0; \varrho)]^{1/2} \leqq L(e^h; x_0, y_0; \varrho) = L(u; x_0, y_0; \varrho).$$

3.27. The preceding proof depends upon analytic functions of a complex variable and therefore it does not apply in the case of three or more independent variables. It is not known at present whether an analogous theorem holds in the case of more than two variables. Clearly, the theorems of 3.25 and 3.26 are contributions to the problem of characterizing subharmonic properties in terms of conditions of the form

$$[A(u^\alpha; x_0, y_0; \varrho)]^{1/\alpha} \leqq [L(u^\beta; x_0, y_0; \varrho)]^{1/\beta}.$$

While the analogous problem for convex functions of a single variable was completely discussed (RADÓ [3]), the general problem for two or more variables seems to present serious difficulties.

3.28. If u is the limit of a sequence u_n of subharmonic functions, then u is also subharmonic if either $u_n \to u$ uniformly (see 3.3) or if $u_n \searrow u$ (see 3.6). MAZURKIEWICZ raised the problem of characterizing those functions which are limits of subharmonic functions in the sense of *convergence in the mean*. This problem was solved by SZPILRAJN [1], whose work we shall review presently.

3.29. A sequence of functions f_n converges *in the mean* to a function f in a domain G if for every domain G', comprised in G together with its boundary, the function f_n is defined and summable in G' for large n and $\iint\limits_{G'} |f - f_n| \to 0$ for $n \to \infty$.

3.30. According to SZPILRAJN, a function u is almost subharmonic in a domain G if it satisfies the following conditions. a) $\dot{u}$ is summable on every measurable set completely interior to G. b) With the possible exception of a set of measure zero, we have $u(x_0, y_0) \leqq A(u; x_0, y_0; r)$ for every point (x_0, y_0) in G and for every r such that the circle $C(x_0, y_0; r)$ is comprised in G together with its interior.

3.31. If a sequence u_n of subharmonic functions converges in the mean to a function u in a domain G, then there exists in G a subharmonic function u^* such that $u^* = u$ almost everywhere in G. Proof. Take two domains G', G'' with boundaries B', B'' such that $G' + B' \subset G''$ and $G'' + B'' \subset G$. We have then an $r_0 > 0$ such that for every point (x_0, y_0) in G' and for $r < r_0$ the circle $C(x_0, y_0; r)$ is comprised in G'' together with its interior. Consider in G' the functions $A_r(x; y; u)$ and $A_r(x, y; u_n)$ (see 2.19). For fixed $r < r_0$ and for large n we have in G'

$$|A_r(x, y; u) - A_r(x, y; u_n)| \leqq \frac{1}{r^2 \pi} \iint\limits_{\xi^2 + \eta^2 < r^2} |u(x+\xi, y+\eta) - u_n(x+\xi, y+\eta)| \, d\xi \, d\eta$$

$$\leqq \frac{1}{r^2 \pi} \iint\limits_{G''} |u - u_n|.$$

The last integral converges to zero for fixed r and for $n \to \infty$, since u_n converges in the mean to u. Hence, for fixed $r < r_0$, $A_r(x, y; u_n)$ converges to $A_r(x, y; u)$ uniformly in G'. But $A_r(x, y; u_n)$ is subharmonic in G' (see 2. 19). Hence (see 3. 3) $A_r(x, y; u)$ is also subharmonic in G'. Finally, we have $A_{r_1}(x, y; u_n) \leqq A_{r_2}(x, y; u_n)$ for $r_1 < r_2$ by 2. 9 and hence also $A_{r_1}(x, y; u) \leqq A_{r_2}(x, y; u)$. Consider now the sequence $u_k^*(x, y) = A_{1/k}(x, y; u)$. By what precedes, this sequence has the following properties. If G' is any domain completely interior to G, then for large k the functions u_k^*, u_{k+1}^*, $\ldots$ are defined and subharmonic in G and $u_k^* \geqq u_{k+1}^* \geqq \cdots$ in G'. Hence, by 3. 6, either $u_k^* \to -\infty$ in G or u_k^* converges in G to a subharmonic function u^*. On the other hand, $u_k^* \to u$ almost everywhere in G by a well-known theorem of LEBESGUE. It follows that $\lim u_k^* = u^*$ is subharmonic in G and $u^* = u$ almost everywhere in G.

3. 32. If u is almost subharmonic in a domain G, then u is the limit of subharmonic functions in the sense of convergence in the mean. Proof. Consider again a pair of subdomains G', G'' as in 3. 31. For small fixed r the function $A_r(x, y; u) = A(u; x, y; r)$ is then continuous in $G'' + B''$ and we have $u(x, y) \leqq A_r(x, y; u)$ almost everywhere in $G'' + B''$, by the definition of an almost subharmonic function. As a consequence, the theorem of TONELLI, referred to in 1. 4, permits us to change the order of integrations necessary to show, starting with the inequality $u(x, y) \leqq A_r(x, y; u)$, that $A_r(x, y; u)$ is subharmonic. Finally, as it is well known, $A_r(x, y; u)$ converges in the mean to $u(x, y)$ for $r \to 0$ (see for instance MORREY [1], p. 687). Hence, if we put $u_n(x, y) = A_{1/n}(x, y; u)$, then the function u_n is subharmonic and converges in the mean to u, and the theorem is proved.

3. 33. Combining 3. 30, 3. 31, 3. 32 we obtain the following theorems (SZPILRAJN [1]). A function u is almost subharmonic in a domain G if and only if there exists in G a subharmonic function u^* such that $u = u^*$ almost everywhere in G. — A function u is almost subharmonic in a domain G if and only if it is the limit, in the sense of convergence in the mean, of some sequence of subharmonic functions.

3. 34. Suppose that a function u satisfies in a domain G condition a) in 3. 30 and that instead of condition b) in 3. 30 it satisfies the following weaker condition b*): With the possible exception of a set of measure zero, there exists for every point (x_0, y_0) in G a $\varrho(x_0, y_0) > 0$ such that $u(x_0, y_0) \leqq A(u; x_0, y_0; r)$ for $r < \varrho(x_0, y_0)$. Then it does *not* follow that u is almost subharmonic in G (SZPILRAJN [1]). Example:

$$u(x, y) = \begin{cases} \log \dfrac{1}{x^2 + y^2} & \text{for} \quad x^2 + y^2 > 0. \\ 0 \quad \text{for} \quad x^2 + y^2 = 0. \end{cases}$$

3. 35. We have the following corollary to 3. 33. If u is almost subharmonic in a domain G, then there exists *exactly one* subharmonic function u^* such that $u = u^*$ almost everywhere in G (SZPILRAJN [1]). This follows immediately from 2. 10.

3. 36. It follows immediately from the definition of an almost subharmonic function (see 3. 30) that a *continuous* almost subharmonic function is subharmonic (SZPILRAJN [1]).

3. 37. As the reviewer could not find in the literature explicit applications of almost subharmonic functions, he takes the liberty to call attention to the following fact. Let u_n denote an *increasing* sequence of subharmonic functions in a domain G, such that there exists a finite constant M for which $\int \int u_n < M$, $n = 1, 2, \ldots$, the integral being taken over G. By 1. 4 the function $u = \lim u_n$ is then summable in G and condition b) in 3. 30 is satisfied by u for every point of G. That is, *the limit of an increasing sequence of subharmonic functions is almost subharmonic as soon as it is summable.* In particular, there exists then a subharmonic function which differs from u at most at the points of a set of measure zero.

Various important problems lead to increasing sequences of subharmonic functions. We mention only the study of *the sweeping-out process* in Potential Theory (see for instance EVANS [4], part II) and the study of the convergence properties of power series of several complex variables (see MONTEL [2], pp. 56—60 and the remarks in RIESZ [4], p. 90 concerning HARTOGS [1]). It seems that the use of almost subharmonic functions might be of advantage in such cases. In a general way, the class of almost subharmonic functions presents the advantage of being closed under a considerably larger number of operations than the class of subharmonic functions.

Chapter IV.

Examples of subharmonic functions.

4. 1. If $u(x, y)$ is a solution of a differential equation of the form $\Delta u = P$, where P is a function of $x, y, u, u_x, \ldots$ etc., then u is subharmonic in every domain G in which P is ≥ 0 (see 3. 2). For various inferences from this remark see BRELOT ([1], pp. 52—55).

4. 2. If h is harmonic in a domain G, then $-h$ is also harmonic there, and by 3. 4 the functions $\overset{+}{h} = \overline{h, 0}$ and $|h| = \overline{h, -h}$ are subharmonic. For $\alpha \geq 1$ the function $|h|^\alpha$ is also subharmonic by 3. 23. More generally, if $h_1, \ldots, h_n$ are harmonic, then for $\alpha \geq 1$ the function $u = (|h_1|^\alpha + \cdots + |h_n|^\alpha)^{1/\alpha}$ is subharmonic, on account of 3. 23. If $f(z) = h_1 + i h_2$, $z = x + iy$, is an analytic function of the complex variable z, then $|f| = (h_1^2 + h_2^2)^{1/2}$ is subharmonic by the preceding remark.

4. 3. As a matter of fact, $|f|$ is of class PL (see 2. 12). This follows immediately from the remark that $\log|f|$ is harmonic at points where $f \neq 0$ and that the limit of $\log|f|$ is equal to $-\infty$ at points where $f = 0$. The far-reaching implications of the preceding facts were first emphasized by F. RIESZ [1, 3, 4, 5], whose papers contain also a number of historical references. Subsequent papers by various authors (practically all the papers quoted in this report) contain many important applications of subharmonic functions in the theory of analytic functions of a complex variable. MONTEL [2] and PRIVALOFF [3, 4] give particularly detailed presentations.

4. 4. If $f = h_1 + i h_2$ is an analytic function of the complex variable $z = x + iy$, then h_1, h_2 are called conjugate harmonic functions. The CAUCHY-RIEMANN equations yield the relations $h_{1x}^2 + h_{2x}^2 = h_{1y}^2 + h_{2y}^2$, $h_{1x} h_{2x} + h_{1y} h_{2y} = 0$ for pairs of conjugate harmonic functions. As a generalization, three functions $u(x, y)$, $v(x, y)$, $w(x, y)$ are said to form *a triple of conjugate harmonic functions* (BECKENBACH and RADÓ [1]) if the following conditions are satisfied. 1) u, v, w are harmonic. 2) $E = G$, $F = 0$, where $E = u_x^2 + v_x^2 + w_x^2$, $G = u_y^2 + v_y^2 + w_y^2$, $F = u_x u_y + v_x v_y + w_x w_y$. According to a theorem of WEIERSTRASS, the surface represented by the equations $X = u(x, y)$, $Y = v(x, y)$, $Z = w(x, y)$ is then a minimal surface (X, Y, Z are CARTESIAN coordinates), and conversely every minimal surface can be represented in this form. As a generalization of 4. 3 we have the following theorem (BECKENBACH and RADÓ [1], p. 653). If u, v, w form a triple of conjugate harmonic functions, then $(u^2 + v^2 + w^2)^{1/2}$ is of class PL (see 2. 12). While the converse is false, it is true that if u, v, w are continuous in a domain G and if $[(u + a)^2 + (v + b)^2 + (w + c)^2]^{1/2}$ is of class PL for every choice of the constants a, b, c, then u, v, w form a triple of conjugate harmonic functions (BECKENBACH and RADÓ [1], p. 654). The first theorem follows by an elementary discussion of the explicit expression for $\Delta \log (u^2 + v^2 + w^2)^{1/2}$. To prove the second theorem, observe that since $[(u + a)^2 + (v + b)^2 + (w + c)^2]^{1/2}$ is of class PL, the function $f = (u + a)^2 + (v + b)^2 + (w + c)^2$ is subharmonic by 3. 24. Let $C(x_0, y_0; r)$ be any circle comprised in G together with its interior. We have then $f(x_0, y_0) \leqq L(f; x_0, y_0; r)$. After some computation, this inequality leads to

$$0 \leqq u(x_0, y_0)^2 + v(x_0, y_0)^2 + w(x_0, y_0)^2 - L(u^2 + v^2 + w^2; x_0, y_0; r)$$

$$- 2a[L(u; x_0, y_0; r) - u(x_0, y_0)] - \cdots - 2c[L(w; x_0, y_0; r) - w(x_0, y_0)].$$

Clearly, if a linear function of a, b, c has a constant sign, the coefficients of a, b, c must vanish. Hence $u(x_0, y_0) = L(u; x_0, y_0; r)$, and thus u is harmonic (by the so-called converse of GAUSS' theorem; see KELLOGG [1], p. 224 or combine 1. 1 and 2. 3). Similarly v and w are harmonic. Now that the derivatives of u, v, w are available, the relations $E = G$, $F = 0$ (cf. 4. 4) can be proved by an elementary discussion of the inequality $\Delta \log f \geqq 0$.

4. 5. Subharmonic functions are related to *surfaces of negative Gaussian curvature* as follows. Let a surface S be given, in terms of CARTESian coordinates X, Y, Z, by equations of the form $X = u(x, y)$, $Y = v(x, y)$, $Z = w(x, y)$. Put $E = u_x^2 + v_x^2 + w_x^2$, $G = u_y^2 + v_y^2 + w_y^2$, $F = u_x u_y + v_x v_y + w_x w_y$. Then the GAUSSian curvature K can be expressed, as it is well known, in terms of E, F, G. Suppose that $E = G$, $F = 0$ (that is, the surface S is given in terms of isothermic parameters). If we put $E = G = \lambda$, then the expression for K in terms of E, F, G reduces to $K = -(1/2\lambda) \Delta \log \lambda$. As $\lambda > 0$, it follows that $K \leqq 0$ if and only if $\Delta \log \lambda \geqq 0$, that is if and only if λ is of class PL (see 2. 12). See BECKENBACH and RADÓ [2] for various geometrical consequences of this relationship between subharmonic functions and surfaces of negative curvature.

Suppose now only that λ is subharmonic. Then it does *not* follow that the GAUSSian curvature of the surface is $\leqq 0$. If however λ is subharmonic for *every* representation of the surface in terms of isothermic parameters, then the GAUSSian curvature of the surface is $\leqq 0$ (BECKENBACH [1]). This follows immediately from 2. 13. This theorem has various interesting geometrical implications (see BECKENBACH [1]).

4. 6. We shall consider presently subharmonic functions arising in Potential Theory (RIESZ [4, 5] and EVANS [4]). We shall use the general notion of a *positive mass-distribution* (RADON [1]). Let us first recall some properties of the class (B) of point-sets which are measurable in the sense of BOREL (BOREL [1], HAUSDORFF [1], KURATOWSKI [1]). The class (B) can be characterized as the *smallest* one of all classes K with the following properties. a) Every closed set belongs to K. b) If $S_1, S_2, \ldots, S_n, \ldots$ is a finite or infinite sequence of sets belonging to K, then $S_1 + S_2 + \cdots$ and $S_1 S_2 \cdots$ also belong to K. It follows easily that if a set S belongs to (B), then the complement of S (the set of points not in S) also belongs to (B). It is then immediate that the class (B) can be also characterized as the smallest one of all classes K^* with the following properties. $\alpha)$ K^* contains every set defined by two relations of the form $x_1 \leqq x < x_2$, $y_1 \leqq y < y_2$. $\beta)$ If S_1, S_2 belong to K^*, then $S_1 S_2$ also belongs to K^*. $\gamma)$ If S_2 and $S_1 \subset S_2$ belong to K^*, then $S_2 - S_1$ also belongs to K^*. $\delta)$ If $S_1, S_2, \ldots, S_n, \ldots$ is a finite or infinite sequence of *non-overlapping* sets belonging to K^*, then $S_1 + S_2 + \cdots$ also belongs to K^*. The equivalence of these two definitions of the class (B) has the following consequence. Denote by $(\mathfrak{P})$ the class of all sets which possess a certain property $\mathfrak{P}$. If it can be shown that $(\mathfrak{P})$ satisfies the conditions $\alpha), \beta), \gamma), \delta)$, then we can assert that *every* set of class (B) possesses the property $\mathfrak{P}$.

4. 7. *Positive mass-distributions.* In the sequel the letters E, e (with subscripts if necessary) will *always* refer to sets of class (B).

Given a bounded set E, let there be assigned to every subset e of E (including the empty set and also E itself) a finite real number $\mu(e)$ such that the following conditions are satisfied. 1) $\mu(e) \geqq 0$. 2) If $e_1, e_2, \ldots$ is a finite or infinite sequence of *non-overlapping* subsets of E, then $\mu(e_1 + e_2 + \cdots) = \mu(e_1) + \mu(e_2) + \cdots$. 3) $\mu(0) = 0$, where $\mu(0)$ denotes the number assigned to the empty set. These conditions being satisfied, $\mu(e)$ is called a positive mass-distribution on E. That is, a positive mass-distribution $\mu(e)$ is a non-negative (and hence monotonic) absolutely additive set-function in the sense of RADON [1]. The theory of these set-functions has been developed to a high degree of efficiency by RADON and it seems that they are generally accepted tools in dealing with problems in Potential Theory. We shall list presently a few facts concerning positive mass-distributions which will be used in the sequel.

4. 8. Let E^* be a set containing the set E on which $\mu(e)$ is defined. Define, for subsets e^* of E^*, $\mu^*(e^*) = \mu(e^*E)$ (a product of two or more sets denotes the set of their common points). Clearly, $\mu^*(e^*)$ is a positive mass-distribution on E^*. If $e^* \subset E$, then $\mu^*(e^*) = \mu(e^*)$, and if $e^*E = 0$ then $\mu^*(e^*) = 0$. Roughly speaking, μ^* vanishes outside of E and μ^* is equal to μ on E. Using this remark, we can always assume that μ is defined on some set E of a convenient type (the interior of a large circle, for instance). RADON [1] assumes that μ is defined on an interval given by relations of the form $x_1 \leqq x < x_2$, $y_1 \leqq y < y_2$. While such assumptions simplify the presentation of the proofs, for the applications it is more convenient to state the theorems for a general set E of class (B).

4. 9. If $e_1 \subset e_2$, then clearly $\mu(e_1) \leqq \mu(e_2)$.

4. 10. If $e_1 \subset e_2 \subset \cdots$ and $e = e_1 + e_2 + \cdots$, then clearly $\mu(e_n) \to \mu(e)$.

4. 11. Given a subset e of E, and an $\varepsilon > 0$, we have a *closed* subset e' of e such that $\mu(e) - \mu(e') < \varepsilon$ (RADON [1], pp. 1313—1314). That is, $\mu(e)$ is the least upper bound of $\mu(e')$ for all *closed* subsets e' of e. This fundamental property is proved on the basis of the remark at the end of 4. 6.

4. 12. Given a positive mass-distribution $\mu(e)$ on E, it might be possible to extend the definition of $\mu(e)$ to a class K^* of subsets of E in such a way that the properties 1), 2), 3) in 4.7 and also the property expressed by the theorem of 4.11 remain valid for all the sets of the class K^*. RADON [1] shows that in a certain sense there exists a largest class K^* satisfying these conditions, and he calls this class K^* the natural range of definition for $\mu(e)$. For instance, if $\mu(e)$ is the LEBESGUE measure of e, then the natural range of definition consists of all sets measurable in the sense of LEBESGUE. The theorem of 4. 11 expresses the important fact that the natural range of definition always includes all sets of class (B). In other words, the class (B) is large enough to possess the closure properties listed in 4. 6 and is also small enough to make valid the theorem of 4. 11. *As stated in 4. 7, we consider only sets of class (B) in connection with positive mass-distributions.*

4. 13. Let $\mu_1(e)$ and $\mu_2(e)$ be given on an *open* set E. If $\mu_1(e)$ and $\mu_2(e)$ have the same value for every *open* subset of E, then $\mu_1 \equiv \mu_2$ on E. Observe that μ_1, μ_2 have then the same value for all *closed* subsets also, and apply 4. 11.

4. 14. Let us denote generally by s and b the set of the interior points and of the boundary points respectively of a square. Then we have the following corollary to 4. 13. If $\mu_1(e)$ and $\mu_2(e)$ are given on an *open* set E, and if $\mu_1(s) = \mu_2(s)$ whenever $s + b$ is comprised in E, then $\mu_1 \equiv \mu_2$ on E. This may be seen as follows. Denote by $S(\xi)$ the set of those points of E which are located on the line $x = \xi$. If $\xi_1, \ldots, \xi_n$ are distinct, then we have $\mu_1(S(\xi_1)) + \cdots + \mu_1(S(\xi_n))$ $\leq \mu(E)$. It follows immediately that $\mu_1(S(\xi)) = 0$, except possibly for a denumerable set of ξ-values. The same holds for $\mu_2(S(\xi))$, and we have a similar statement in terms of the y-coordinate. It follows that we have a point (x_0, y_0) with the following property. Denote by D_n the subdivision of the plane by means of the lines $x = x_0 + k/2^n$, $y = y_0 + j/2^n$, $k, j = 0, \pm 1, \ldots$ Let $s + b$ be any closed square of D_n. Then $\mu_1(E b) = \mu_2(E b) = 0$. On account of 4. 9, 4. 10 and 4. 13 the theorem follows now immediately if we approximate the open subsets of E by squares taken from D_n.

4. 15. *The Stieltjes-Radon integral* (RADON [1], p. 1322). Let $u(e)$ be a positive mass-distribution given on E. Denote by Q a variable point of E with coordinates (ξ, η). We shall write $Q = (\xi, \eta)$ in the sense that we shall use whichever of the notations Q and (ξ, η) will be more convenient. Let $f(Q) = f(\xi, \eta)$ be a function which is *uniformly continuous* on E. Subdivide E into a finite number of nonoverlapping subsets $e_1, \ldots, e_n$. Denote by δ_k the diameter of e_k (that is, the least upper bound of the distances of pairs of points in e_k) and by δ the largest one of $\delta_1, \ldots, \delta_n$. We shall say that $e_1, \ldots, e_n$ form a subdivision D of E with norm δ. Pick a point $Q_k = (\xi_k, \eta_k)$ in e_k, $k = 1, \ldots, n$, and form the sum $\Sigma = \Sigma f(Q_k) \mu(e_k)$, $k = 1, \ldots, n$. In exactly the same way as in the case of the RIEMANN integral, it follows that the sum Σ approaches a limit, depending only upon f and μ, if the norm of the subdivision approaches zero. This limit is the STIELTJES-RADON integral $\int_E f(Q)\, d\mu(e_Q)$. The symbol e_Q is used to avoid misunderstandings in case f depends upon further variables.

4. 16. Note that the STIELTJES-RADON integral is defined, for the time being, only for functions which are *uniformly contiuuous* on E. We shall consider later one of the various possible generalizations. It seems unnecessary to state explicitly all the simple properties of the STIELTJES-RADON integral which will be used in the sequel. As an example, we mention the following property. Suppose that E_1 is a subset of E, such that $\mu(E_1) = 0$. Put $E_2 = E - E_1$. Then $\int_E f(Q)\, d\mu(e_Q) = \int_{E_2} f(Q)\, d\mu(e_Q)$. This becomes obvious if we observe that we can subdivide E_1 and E_2 separately and that $\mu(E_1) = 0$ implies $\mu(e_1) = 0$ for every subset e_1 of E_1.

4. 17. An important special case of positive mass-distributions is obtained as follows. Let $w(Q) = w(\xi, \eta)$ be a non-negative summable function on E. For convenience, we shall use notations like $\int\int w(\xi, \eta)\, d\xi\, d\eta = \int\int w(Q)\, da_Q$, where the symbol da_Q is to remind us of the variable of integration $Q = (\xi, \eta)$, of the area-element $d\xi\, d\eta$, and also of the fact that we are dealing with a LEBESGUE integral, in contradistinction with STIELTJES-RADON integrals. Consider now the set-function $\mu(e) = \int\int_e w(Q)\, da_Q$ on E. Obviously $\mu(e)$ is a positive mass-distribution on E. We have, for every function $f(Q)$ which is *uniformly continuous* on E, the relation $\int_E f(Q)\, d\mu(e_Q) = \int\int_E f(Q)\, w(Q)\, da_Q$.

To see this, take a subdivision $e_1, \ldots, e_n$ of E, and use the uniform continuity of f in comparing the sums $\Sigma f(Q_k)\, \mu(e_k)$ and

$$\Sigma \int_{e_k} f(Q)\, w(Q)\, da_Q = \int_E f(Q)\, w(Q)\, da_Q.$$

4. 18. Given the positive mass-distribution $\mu(e)$ on E, we have to define the integral (potential of the negative mass-distribution $-\mu(e)$)

$$-\int_E \log \frac{1}{PQ}\, d\mu(e_Q).$$

The symbol PQ denotes the distance of the points $P = (x, y)$ and $Q = (\xi, \eta)$, where Q varies on E and P varies in the whole plane. The existence and the properties of this potential will be discussed presently (cf. RIESZ [5], part II; note that RIESZ uses a somewhat different definition of positive mass-distributions).

4. 19. Let us put

$$l(P, Q) = l(x, y; \xi, \eta) = \begin{cases} -\log(1/PQ) & \text{for} \quad P \neq Q, \\ -\infty & \text{for} \quad P = Q. \end{cases}$$

For fixed Q, $l(P, Q)$ is clearly a subharmonic function of P and conversely. For fixed Q and $P \neq Q$, $l(P, Q)$ is a harmonic function of P and conversely.

4. 20. Put, for $\sigma > 0$,

$$l^{(\sigma)}(P, Q) = l^{(\sigma)}(x, y; \xi, \eta) = \begin{cases} l(P, Q) & \text{for} \quad PQ \geqq \sigma, \\ -\log(1/\sigma) & \text{for} \quad PQ \leqq \sigma. \end{cases}$$

For fixed Q, $l^{(\sigma)}(P, Q)$ is a subharmonic function of P by 3. 4, and conversely. Clearly, $l^{(\sigma)}(P, Q)$ is a continuous function of P and Q, and its continuity is uniform if P and Q vary on bounded sets. Also $l^{(\sigma)}(P, Q) \searrow l(P, Q)$ for $\sigma \searrow 0$.

4. 21. Put, for $r > 0$,

$$l_r(P, Q) = l_r(x, y; \xi, \eta) = \frac{1}{2\pi} \int_0^{2\pi} l(x + r\cos\varphi, y + r\sin\varphi; \xi, \eta) \, d\varphi.$$

We find by direct elementary computation the formula $l_r(P, Q) = l^{(r)}(P, Q)$ (see 4. 20).

4. 22. Put, for $r > 0$ and $\sigma > 0$,

$$l_r^{(\sigma)}(P, Q) = l_r^{(\sigma)}(x, y; \xi, \eta) = \frac{1}{2\pi} \int_0^{2\pi} l^{(\sigma)}(x + r\cos\varphi, y + r\sin\varphi; \xi, \eta) \, d\varphi.$$

By 4. 20 and 4. 21 we have then $l_r^{(\sigma)}(P, Q) \searrow l_r(P, Q)$ for $\sigma \searrow 0$. Since $l_r^{(\sigma)}(P, Q)$ and $l_r(P, Q)$ are continuous, it follows by a well-known theorem of DINI (see for instance PÓLYA-SZEGŐ [1], p. 225, problem 126) that $l_r^{(\sigma)}(P, Q) \to l_r(P, Q)$ for $\sigma \to 0$ *uniformly* if P and Q vary on bounded sets (since such sets are comprised in bounded *closed* sets). By 4. 19 to 4. 21 we have for $r + \sigma < PQ$ the formula $l_r^{(\sigma)}(P, Q) = l(P, Q)$.

4. 23. We define now (cf. RIESZ [5], part II and DANIELL [1])

$$u(P) = \int_E l(P, Q) \, d\mu(e_Q) = -\int_E \log\frac{1}{PQ} \, d\mu(e_Q) = \lim_{\sigma \to 0} u^{(\sigma)}(P),$$

where, for $\sigma > 0$,

$$u^{(\sigma)}(P) = \int_E l^{(\sigma)}(P, Q) \, d\mu(e_Q).$$

Since $l^{(\sigma)}(P, Q)$ is uniformly continuous if P and Q vary on bounded sets, $u^{(\sigma)}(P)$ is a well-defined and continuous function of P in the whole plane. From $\mu \geqq 0$ it follows that $u^{(\sigma)}(P)$ decreases if $\sigma > 0$ decreases, and thus $u(P) = \lim u^{(\sigma)}(P)$ exists for every P, the value of $u(P)$ being possibly equal to $-\infty$. We have $-\infty \leqq u(P) < +\infty$, $u^{(\sigma)}(P) \searrow u(P)$ for $\sigma \searrow 0$, and we can assert also that $u(P)$ is upper semi-continuous, since we obtained $u(P)$ as the limit of a decreasing sequence of continuous functions. We shall see now that $u(P)$ is subharmonic.

4. 24. Put, for $\sigma > 0$ and $r > 0$,

$$u_r^{(\sigma)}(P) = u_r^{(\sigma)}(x, y) = \frac{1}{2\pi} \int_0^{2\pi} u^{(\sigma)}(x + r\cos\varphi, y + r\sin\varphi) \, d\varphi.$$

Since $l^{(\sigma)}(P, Q)$ is continuous (see 4. 20, 4. 23), we obtain by an obviously permissible change in the order of integrations the formula $u_r^{(\sigma)}(P) = \int_E l_r^{(\sigma)}(P, Q) \, d\mu(e_Q)$. On account of 4. 22, $l_r^{(\sigma)}(P, Q) \to l_r(P, Q)$ uniformly for $\sigma \to 0$. Hence $u_r^{(\sigma)}(P) \to \int_E l_r(P, Q) \, d\mu(e_Q)$ for $\sigma \to 0$.

4. 25. Since $l^{(\sigma)}(P, Q)$ is a subharmonic function of P (see 4. 20) we have, by 2. 3, $l_r^{(\sigma)}(P, Q) \geq l^{(\sigma)}(P, Q)$. Hence (see 4. 24), $u_r^{(\sigma)}(P) \geq u^{(\sigma)}(P)$. Thus $u^{(\sigma)}(P)$ is subharmonic by 2. 3. Suppose now that the point P has a positive distance δ from the set E. For $\sigma < \delta/2$, $r < \delta/2$ we have then by 4. 22 and 4. 24

$$u_r^{(\sigma)}(P) = \int_E l_r^{(\sigma)}(P, Q) \, d\mu(e_Q) = \int_E l(P, Q) \, d\mu(e_Q) = \int_E l^{(\sigma)}(P, Q) \, d\mu(e_Q) = u^{(\sigma)}(P).$$

Consider then an open set O such that $PQ > \delta > 0$ for P in O and Q in E. By what precedes, we have for P in O, $r < \delta/2$, $\sigma < \delta/2$ the relations $u_r^{(\sigma)}(P) = u^{(\sigma)}(P)$ and $u^{(\sigma)}(P) = \int_E l(P, Q) \, d\mu(e_Q)$. The first relation shows that $u^{(\sigma)}(P)$ is harmonic in O (converse of GAUSS' theorem, see KELLOGG [1], p. 224). The second relation shows that $u(P) = \lim u^{(\sigma)}(P) = u^{(\sigma)}(P)$ in O for $\sigma < \delta/2$. Thus $u(P)$ is also harmonic in O.

4. 26. As $u(P)$ is the limit of a decreasing sequence of subharmonic functions $u^{(\sigma)}(P)$, it follows by 3. 6 that either $u \equiv -\infty$ in the whole plane or u is subharmonic in the whole plane. The first case is excluded by the remark that u is harmonic outside of a sufficiently large circle (see 4. 25). Consequently *the potential*

$$u(P) = -\int_E \log \frac{1}{PQ} \, d\mu(e_Q)$$

is subharmonic in the whole plane (RIESZ [5], part II). In particular (see 1. 10), $u > -\infty$ almost everywhere.

4. 27. Let E_1 be an *open* subset of E, such that $\mu(E_1) = 0$. Put $E_2 = E - E_1$. We have then (see 4. 16)

$$u^{(\sigma)}(P) = \int_E l^{(\sigma)}(P, Q) \, d\mu(e_Q) = \int_{E_2} l^{(\sigma)}(P, Q) \, d\mu(e_Q),$$

and consequently (see 4. 23)

$$u(P) = \int_{E_2} l(P, Q) \, d\mu(e_Q).$$

Hence, on account of the remark at the end of 4. 25, $u(P)$ is harmonic in E_1. Summing up: *the potential $u(P)$ is harmonic on every open set which contains no mass.*

4. 28. As $u^{(\sigma)} \searrow u$ for $\sigma \searrow 0$, we have (see 1. 4, 4. 24, 4. 21, 4. 23)

$$L(u; x, y; r) = \lim_{\sigma \to 0} L(u^{(\sigma)}; x, y; r) = \lim_{\sigma \to 0} u_r^{(\sigma)}(P) = \int_E l_r(P, Q)\, d\mu(e_Q)$$

$$= \int_E l^{(r)}(P, Q)\, d\mu(e_Q) = u^{(r)}(x, y).$$

These formulas throw a new light on the definition of the potential $u(P)$ given in 4. 23.

4. 29. Take a point $P_0 = (x_0, y_0)$ and choose r so large that the set E is completely interior to the circle $C(x_0, y_0; r)$. For Q in E we have then $P_0 Q < r$ and hence $l_r(P_0, Q) = \log r$ (see 4. 21). It follows then by 4. 28 that $\mu(E) \log r = L(u; x_0, y_0; r)$.

4. 30. On $C(x_0, y_0; r)$ we can use the derivatives of u, since u is harmonic there by 4. 27. Let us write C_r for $C(x_0, y_0; r)$. We have then (on account of 4. 29) the formula

$$\frac{1}{2\pi} \int_{C_r} \frac{\partial u}{\partial n_e}\, ds = r \frac{d}{dr} L(u; x_0, y_0; r) = \mu(E).$$

4. 31. Consider now any smooth JORDAN curve Γ, such that the set E is completely interior to Γ. We can choose then the circle $C(x_0, y_0; r)$ of 4. 29 in such a way that Γ is enclosed by $C(x_0, y_0; r)$. As u is harmonic between and on these two curves, the line integral of 4. 30 has the same value for both curves. Hence, the total mass $\mu(E)$ can be expressed in terms of the potential u by the familiar formula

$$\mu(E) = \frac{1}{2\pi} \int_{\Gamma} \frac{\partial u}{\partial n_e}\, ds,$$

where Γ is any smooth JORDAN curve such that E is completely interior to Γ (RIESZ [5], part II).

4. 32. The problem of expressing the mass $\mu(e)$, for every subset of class (B), in terms of the potential u will be considered in Chapter V.

4. 33. Let $w(Q) = w(\xi, \eta)$ be a non-negative summable function on a bounded set E of class (B). Consider the function (see 4. 17 for notations)

$$v(P) = -\iint_E \log \frac{1}{PQ}\, w(Q)\, da_Q.$$

This integral can be interpreted in various ways. One usual interpretation is expressed by the formula (cf. 4. 20)

$$v(P) = \lim_{\sigma \to 0} \iint_E l^{(\sigma)}(P, Q)\, w(Q)\, da_Q.$$

By 4. 17 and 4. 23 it follows that

$$v(P) = -\int_E \log \frac{1}{PQ}\, d\mu(e_Q),.$$

where the positive mass-distribution $\mu(e)$ is defined by $\mu(e) = \int_e w(T)\, da_T$.

That is, the potentials of mass-distributions with a summable negative density $-w(\xi, \eta)$, as considered in Potential Theory, are included among the potentials of general negative mass-distributions $-\mu(e)$.

4.34. Using the fact that potentials of negative mass-distributions are subharmonic, it is possible to construct subharmonic functions with various types of discontinuities (RIESZ [5], part I, p. 336; BRELOT [1], pp. 42—47; EVANS [5], p. 421). These examples show the great variety of new possibilities as compared with the one-dimensional case of convex functions of a single variable.

Chapter V.

Harmonic majorants of subharmonic functions.

5.1. Throughout this Chapter, u will denote a function which is subharmonic in a domain G. Consider a region $G' + B'$ comprised in G and a function H which is continuous in $G' + B'$ and harmonic in G'. If $H \geq u$ on B', then $H \geq u$ in G' also, by the definition of a subharmonic function. Naturally, one will try to use a harmonic majorant H which is as small as possible. Suppose that $G' + B'$ is a DIRICHLET region and suppose also that u is *continuous*. The solution H of the DIRICHLET problem for G' with the boundary condition $H = u$ on B' is then obviously the best harmonic majorant in G'. If however u is not continuous, then it is not clear that there exists a harmonic majorant in G' which should be considered the best. This situation lead to investigations which will be reviewed presently.

5.2. Consider a DIRICHLET region $G' + B'$ comprised in G. By 1.3 we have on B' a sequence of continuous functions φ_k such that $\varphi_k \searrow u$ on B'. Denote by H_k the solution of the DIRICHLET problem for G' with the boundary condition $H_k = \varphi_k$ on B'. Then we have (see 1.3) $H_k \geq H_{k+1}$ and $H_k \geq u$ on $G' + B'$ and therefore H_k converges in G' to a harmonic function $\bar{h} \geq u$.

5.3. The function $\bar{h}$ of 5.2 has the following property. Let H be continuous and $\geq u$ in $G' + B'$ and harmonic in G'. Then $H \geq \bar{h}$ in G' (RIESZ [5], part I, p. 334). To see this, give any $\varepsilon > 0$. As $\varphi_k \searrow u < H + \varepsilon$ and as φ_k and H are continuous on the closed set B', it follows by the HEINE-BOREL theorem that we have a $\varkappa = \varkappa(\varepsilon)$ such that $\varphi_k < H + \varepsilon$ on B' for $k > \varkappa$. But $H_k = \varphi_k$ on B' and H_k is harmonic in G'. Hence $H_k < H + \varepsilon$ for $k > \varkappa$. As $\varepsilon > 0$ is arbitrary and $H_k \searrow \bar{h}$ in G', it follows that $\bar{h} \leq H$ in G'.

5.4. The function $\bar{h}$ of 5.2 depends only upon the values of u on B' (RIESZ [5], part I, pp. 333—334). Indeed, take a second sequence φ'_k and denote by H'_k and $\bar{h}'$ the corresponding harmonic functions. By 5.3 we have $\bar{h} \leq H'_k$ and $\bar{h}' \leq H_k$ in G' and the theorem follows for $k \to \infty$.

Consider now two functions u_1, u_2 which are subharmonic in G and equal to each other on B'. As we can use then the same sequence φ_k for both functions, there corresponds the same function $\bar{h}$ to u_1 and to u_2.

The harmonic function $\bar{h}$ defined in 5.2 will be called *the best harmonic majorant* (B. H. M.) of u in G' (RIESZ [5], part I, p. 334). By what precedes, u depends solely upon the values of u on the boundary B' of G'. The term best harmonic majorant suggests various questions which will be considered later in this chapter. It should be noted that a B. H. M. is only defined for DIRICHLET subregions $G' + B'$. If u is continuous on B', then we can use $\varphi_k = u$ as the sequence leading to $\bar{h}$, and it follows that in this special case $\bar{h}$ is simply the solution of the DIRICHLET problem for G' with the boundary condition $\bar{h} = u$ on B'. Another important special case arises if $G' + B'$ is a closed circular disc, while u is a general subharmonic function. The function H_k of 5.2 is then given in G' by the formula of POISSON. As the POISSON kernel is positive and $\varphi_k \searrow u$ on B', we infer from 1.4 that $\bar{h}$ is also given by the formula of POISSON with u itself as the given boundary function. Clearly, a similar remark holds for DIRICHLET subregions with smooth boundaries.

5.5. Consider in G three JORDAN curves C_1, C_2, C_3, each of which is enclosed by the next one to the right, such that the three doubly connected domains D_{12}, D_{13}, D_{23} bounded by these curves are also comprised in G. Denote by $\bar{h}_{12}, \bar{h}_{13}, \bar{h}_{23}$ the B. H. M. of u in D_{12}, D_{13}, D_{23} respectively. Then $\bar{h}_{13} - \bar{h}_{12}$ is non-negative in D_{12} and vanishes continuously on C_1, and $\bar{h}_{13} - \bar{h}_{23}$ is non-negative in D_{23} and vanishes continuously on C_3 (RIESZ [5], part I, p. 341). Proof. Consider $\bar{h}_{13} - \bar{h}_{12}$, for instance. Take a sequence of continuous functions φ_k^i on C_i such that $\varphi_k^i \searrow u$ on C_i, $i = 1, 2, 3$. Denote by H_k^{13} the solution of the DIRICHLET problem for D_{13} with the boundary condition $H_k^{13} = \varphi_k^1$ on C_1, $H_k^{13} = \varphi_k^3$ on C_3, and let H_k^{12} have the same meaning with respect to D_{12}. Finally, denote by C_4 an auxiliary JORDAN curve in D_{12} which encloses C_1. Then the sequence $H_k = H_k^{13} - H_k^{12}$ converges uniformly on the boundary of the domain D_{14} bounded by C_1 and C_4 and hence (see KELLOGG [1], p. 248) this sequence converges *uniformly* in $D_{14} + C_1 + C_4$ to a limit function h which is continuous in $D_{14} + C_1 + C_4$, equal to zero on C_1, and equal to $\bar{h}_{13} - \bar{h}_{12}$ in D_{14}. This proves that $\bar{h}_{13} - \bar{h}_{12}$ vanishes continuously on C_1. By 5.3 we have $H_k^{13} \geq \bar{h}_{12}$ in D_{12}, and for $k \to \infty$ it follows that $\bar{h}_{13} \geq \bar{h}_{12}$ in D_{12}.

If the interior of C_2 is comprised in G, and if we denote be $\bar{h}_2$ the B. H. M. of u in the interior of C_2, then the same reasoning shows that $\bar{h}_2 - \bar{h}_{12}$ is non-negative in D_{12} and vanishes continuously on C_2.

5. 6. Let G' be a bounded domain with boundary B'. Using subdivisions of the plane into congruent squares in a familiar fashion (KELLOGG [1], p. 317), we obtain a sequence of regions $G'_n + B'_n$ which approximate G' in the following sense. a) $G'_n + B'_n \subset G'$. b) $G'_n + B'_n \subset G'_{n+1}$. c) For every *closed* set S in G' there exists an $n_0 = n_0(S)$ such that S is in G'_n for $n > n_0$. d) B'_n consists of a finite number of JORDAN curves as smooth as desired (in particular, $G'_n + B'_n$ is a DIRICHLET region). The following statements are easy consequences of the preceding properties. e) Given $\varepsilon > 0$, denote by S_ε the set of those points in G' whose distance from B' is less than ε. Then for every $\varepsilon > 0$ there exists an $m = m(\varepsilon)$ such that B'_n is comprised in S_ε for $n > m$. f) The area of G'_n converges to the area of G, and consequently the measure of $G' - G'_n$ converges to zero.

5. 7. Given a subharmonic function u in a domain G, consider a domain G' comprised in G. Suppose that there exists a function H_0 which is harmonic and $\geqq u$ in G' (this assumption is clearly satisfied if the boundary of G' is also comprised in G). Then there exists in G' a harmonic function h^* such that 1) $u \leqq h^*$ in G' and 2) every function H which is harmonic and $\geqq u$ in G' is also $\geqq h^*$ (RIESZ [5], part II, p. 358). Proof. Approximate G' by a sequence $G'_n + B'_n$ as described in 5.6. Denote by $\bar{h}_n$ the B. H. M. of u in G'_n. By 5.3 we have $u \leqq \bar{h}_n \leqq \bar{h}_{n+1} \leqq H_0$ in G'_n. It follows then from the theorem of HARNACK that the sequence $\bar{h}_n$ converges in G' to a function h^* which is harmonic in G' and which satisfies there the inequalities $u \leqq h^* \leqq H_0$. By 5.3 we have also $\bar{h}_n \leqq H$ for every function H which is harmonic and $\geqq u$ in G' and for $n \to \infty$ it follows that $h^* \leqq H$ in G'.

5. 8. The harmonic function h^* of 5.7 is obviously unique. It may be called *the least harmonic majorant* (L. H. M.) of u in G' (RIESZ [5], part II, p. 357). If $G' + B'$ is a DIRICHLET region comprised in G, then the B. H. M. $\bar{h}$ and the L. H. M. h^* of u in G' both exist. Clearly $h^* \leqq \bar{h}$. As $\bar{h}$ depends solely upon the values of u on B' and h^* depends solely upon the values of u in G', it is not evident that $\bar{h}$ and h^* should be identical. The identity of $\bar{h}$ and h^* was established for special types of subregions $G' + B'$ by F. RIESZ ([5], part I, p. 334) and by BRELOT ([1], p. 18). We shall see later in this Chapter that $\bar{h}$ and h^* are always identical.

5. 9. The majorants $\bar{h}$ and h^* depend upon u and upon G'. Some aspects of this dependence were investigated by MALCHAIR [2]. In the way of illustration we quote one of his results. Consider in a domain G a uniformly convergent sequence of subharmonic functions u_n.

Then the limit function u is also subharmonic by 3.3. Denote by $G'_n + B'_n$ a sequence of regions which approximate G in the sense of 5.6, and by $\bar{h}_n$ the B.H.M. of u_n in G'_n. Then $\bar{h}_n$ converges to the L.H.M. of u in G provided that this L.H.M. exists. The proof is similar to that in 5.7.

5.10. *A lemma on harmonic functions* (RIESZ [5], part I, p. 341). Consider two JORDAN curves C_1, C_2 such that C_1 is enclosed by C_2, and denote by D the doubly connected domain bounded by these curves. Let h be a function which is continuous on $D + C_1 + C_2$ and harmonic and *non-negative* in D. Take a smooth JORDAN curve Γ in D which encloses C_1. If $h = 0$ on C_1, then $\int\limits_{\Gamma} (\partial h/\partial n_e)\, ds \geqq 0$, and if

$h = 0$ on C_2, then $\int\limits_{\Gamma} (\partial h/\partial n_e)\, ds \leqq 0$. Proof. *Special case.* Suppose that $h = 0$ on C_1, for instance, and suppose that C_1 is sufficiently *smooth*. Then the first and second derivatives of h remain continuous on C_1, and the line integral has the same value for Γ and for C_1 (KELLOGG [1], p. 212). The integral taken on C_1 is however obviously $\geqq 0$. If h vanishes on C_2, and if C_2 is sufficiently smooth, then the theorem is equally obvious. *General case.* Suppose that $h = 0$ on C_1, for instance. Take two *smooth* JORDAN curves C_3, C_4 such that each of the curves $C_3, C_1, \Gamma, C_4, C_2$ is enclosed by the next one to the right (C_3 being close to C_1 and C_4 close to C_2). Denote by H_{34} the solution of the DIRICHLET problem for the domain bounded by C_3 and C_4 with the boundary condition $H_{34} = 0$ on $C_3, H_{34} = h$ on C_4. Apply the *special case* of the theorem to $H_{34} - h$ in the domain between C_1 and C_4, then to H_{34} in the domain between C_3 and C_4, and combine the resulting inequalities. The case when $h = 0$ on C_2 is discussed in a similar manner.

5.11. F. RIESZ ([5], part I) introduced the following quantities in the study of subharmonic functions. Let u be subharmonic in a domain G. Take in G a pair of JORDAN curves C_1, C_2 such that C_1 is enclosed by C_2 and the domain D_{12} between C_1 and C_2 is comprised in G. Denote by $\bar{h}_{12}$ the B.H.M. of u in D_{12} and put

$$F(C_1, C_2; u) = \frac{1}{2\pi} \int\limits_{\Gamma} \frac{\partial \bar{h}_{12}}{\partial n_e}\, ds,$$

where Γ is a smooth JORDAN curve in D_{12} which encloses C_1, and n_e refers to the outward normal of Γ. The quantity $F(C_1, C_2; u)$ is clearly independent of Γ (see KELLOGG [1], p. 212). If u_1, u_2 are both subharmonic in G, then clearly

$$F(C_1, C_2; u_1) + F(C_1, C_2; u_2) = F(C_1, C_2; u_1 + u_2).$$

5.12. If the interior of C_2 is comprised in G, then $F(C_1, C_2; u) \geqq 0$ (RIESZ [5], part I, p. 342). Proof. Denote by D_{12} the domain between C_1 and C_2 and by D_2 the interior of C_2. Let $\bar{h}_{12}$ and $\bar{h}_2$ be the B.H.M. of u in D_{12} and D_2 respectively. Take a smooth JORDAN curve Γ in D_{12} which encloses C_1. Then

$$\frac{1}{2\pi}\int_{\Gamma} \frac{\partial \bar{h}_2}{\partial n_e}\, ds = 0, \quad \frac{1}{2\pi}\int_{\Gamma} \frac{\partial \bar{h}_{12}}{\partial n_e}\, ds = F(C_1, C_2; u),$$

and hence, by 5.5 and 5.10,

$$F(C_1, C_2; u) = -\frac{1}{2\pi}\int_{\Gamma} \frac{\partial(\bar{h}_2 - \bar{h}_{12})}{\partial n_e}\, ds \geqq 0.$$

5.13. Take in G three JORDAN curves C_1, C_2, C_3 such that C_1 is enclosed by C_2, C_2 is enclosed by C_3, and the domains D_{12}, D_{13}, D_{23} bounded by these curves are comprised in G. Then $F(C_1, C_2; u) \leqq F(C_1, C_3; u) \leqq F(C_2, C_3; u)$ (RIESZ [5], part I, p. 340). Proof. Take a smooth JORDAN curve Γ in D_{12} which encloses C_1 and denote by $\bar{h}_{12}$, $\bar{h}_{13}$, $\bar{h}_{23}$ the B.H.M. of u in D_{12}, D_{13}, D_{23} respectively. Then

$$F(C_1, C_3; u) - F(C_1, C_2; u) = \frac{1}{2\pi}\int_{\Gamma} \frac{\partial(\bar{h}_{13} - \bar{h}_{12})}{\partial n_e}\, ds,$$

and this integral is $\geqq 0$ by 5.5 and 5.10. The inequality $F(C_1, C_3; u) \leqq F(C_2, C_3; u)$ is proved in a similar way.

5.14. Consider the particular case when C_1 and C_2 are concentric circles with centre (x_0, y_0) and radii r_1 and $r_2 > r_1$. Denote by $\bar{h}_{12}$ the B.H.M. of u in the domain D_{12} between C_1 and C_2 and by C_r the concentric circle with radius r, $r_1 < r < r_2$. Then (see 1.5 for notations)

$$F(C_1, C_2; u) = r \frac{d}{dr} L(\bar{h}_{12}; x_0, y_0; r).$$

On the other hand, the reasoning used in 1.12 and 1.13 shows that $L(\bar{h}_{12}; x_0, y_0; r) = a \log r + b$, where $a \log r + b$ is the linear function of $\log r$ which is equal to $L(u; x_0, y_0; r_1)$ for $r = r_1$ and to $L(u; x_0, y_0; r_2)$ for $r = r_2$. Combining these relations, we obtain the formula (RIESZ [5], part I, p. 340)

$$F(C_1, C_2; u) = \frac{L(u; x_0, y_0; r_2) - L(u; x_0, y_0; r_1)}{\log r_2 - \log r_1}.$$

5.15. The theorems of 2.4 and 2.5 appear now, on account of 5.14, as special cases of the theorems of 5.12 and 5.13.

5.16. Given a potential (cf. 4.23)

$$u(P) = -\int_{G} \log \frac{1}{PQ}\, d\mu(e_Q),$$

where G is a bounded domain, there arises the problem to express the positive mass-distribution $\mu(e)$ in terms of u. If the distribution $\mu(e)$ is smooth, and if $G' + B'$ is a region in G with smooth boundary B', then the problem is solved by the classical formula (KELLOGG [1], pp. 155—156)

$$\mu(G') = \frac{1}{2\pi}\iint\limits_{G'} \Delta u(x,y)\,dx\,dy = \frac{1}{2\pi}\int\limits_{B'} \frac{\partial u}{\partial n_e}\,ds\,,$$

where n_e refers to the exterior normal with respect to G'. For a general $\mu(e)$ the problem was solved by G. C. EVANS in terms of a certain function of curves (EVANS [1], p. 271 and p. 285). We shall discuss this problem presently in terms of the quantities $F(C_1, C_2; u)$ introduced by F. RIESZ.

5. 17. The potential $u(P)$ of 5. 16 is subharmonic in the whole plane (see 4. 26) and therefore the preceding theorems apply to $u(P)$. Take two JORDAN curves C_1, C_2, such that C_1 is enclosed by C_2, and G is completely interior to C_1. By 4. 27 the potential u is harmonic on and between C_1 and C_2 and hence u is its own best harmonic majorant in the domain between C_1 and C_2. By 4. 31 and 5. 11 we obtain therefore for the total mass $\mu(G)$ the formula $\mu(G) = F(C_1, C_2; u)$. Consider next two JORDAN curves C_1, C_2 such that C_1 is enclosed by C_2 and both curves are comprised in a simply connected subdomain G' of G with $\mu(G') = 0$. Then, by 4. 27, u is harmonic in G' and again u is its own harmonic majorant in the domain between C_1 and C_2. If Γ is a smooth JORDAN curve which encloses C_1 and is enclosed by C_2, then it follows from the preceding remark that

$$F(C_1, C_2; u) = \frac{1}{2\pi}\int\limits_{\Gamma} \frac{\partial u}{\partial n_e}\,ds = 0\,,$$

since u is harmonic in and on Γ (see KELLOGG [1], p. 212).

5. 18. Take now five JORDAN curves $C_1, \ldots, C_5$ such that each one is enclosed by the next one to the right. Denote by D_i the interior of C_i, by D_{ij} the domain between C_i and C_j, and by h_i, h_{ij} the B. H. M. of u in D_i, D_{ij} respectively. We have then $F(C_1, C_2; u) \leq \mu(GD_3) \leq F(C_4, C_5; u)$ (RIESZ [5], part II, pp. 331—335).

5. 19. To prove the preceding theorem, introduce on G the distributions $\mu'(e) = \mu(eGD_3)$, $\mu''(e) = \mu(e(G - GD_3))$, and the corresponding potentials (cf. 4. 16)

$$u'(P) = -\int\limits_{G} \log\frac{1}{PQ}\,d\mu'(e_Q) = -\int\limits_{GD_3} \log\frac{1}{PQ}\,d\mu(e_Q)\,,$$

$$u''(P) = -\int\limits_{G} \log\frac{1}{PQ}\,d\mu''(e_Q) = -\int\limits_{G - GD_3} \log\frac{1}{PQ}\,d\mu(e_Q)\,.$$

Then $\mu'(e) + \mu''(e) = \mu(e)$ and consequently $u'(P) + u''(P) = u(P)$. Hence (see 5. 11) $F(C_1, C_2; u') + F(C_1, C_2; u'') = F(C_1, C_2; u)$. We have, by 5. 17, $F(C_1, C_2; u'') = 0$ and $F(C_4, C_5; u') = \mu'(GD_3) = \mu(GD_3)$. Repeated application of 5. 13 yields $F(C_1, C_2; u') \leq F(C_4, C_5; u')$. The inequality $F(C_1, C_2; u) \leq \mu(GD_3)$ follows by combining these relations. The inequality $\mu(GD_3) \leq F(C_4, C_5; u)$ is obtained in a similar fashion.

5. 20. Consider now a *simply connected* domain G' comprised in G. Take a sequence of pairs of JORDAN curves C'_n, C''_n in G' such that 1) C'_n is enclosed by C''_n and 2) every closed set in G' is comprised in the interior of C'_n for sufficiently large n. Then $F(C'_n, C''_n; u) \to \mu(G')$ (RIESZ [5], part II, p. 336). This follows immediately from 5. 18, 5. 13 and 4. 10.

5. 21. By 5. 20, $\mu(G')$ is determined in terms of u whenever G' is a simply connected subdomain of G. A similar reasoning yields the determination of $\mu(e)$ for multiply connected subdomains (RIESZ [5], part II, p. 336).

5. 22. From 5. 20 and 4. 14 we infer the following theorem. If $\mu_1(e)$, $\mu_2(e)$ are positive mass-distributions on a bounded domain G, and if the corresponding potentials

$$u_1(P) = -\int\limits_G \log \frac{1}{PQ}\, d\mu_1(e_Q)\,, \qquad u_2(P) = -\int\limits_G \log \frac{1}{PQ}\, d\mu_2(e_Q)$$

are equal to each other in G, then $\mu_1(e) \equiv \mu_2(e)$ (remember that we consider only subsets e of class (B)). For the sake of accuracy it should be observed that F. RIESZ ([5], part II) considers positive mass-distributions defined in a somewhat different manner. In particular, his $\mu(e)$ is defined only for *open* sets e. The remark that the results of F. RIESZ include the preceding uniqueness theorem is due to EVANS ([4], part II, p. 203).

5. 23. *A lemma on sequences of harmonic functions* (KELLOGG [1], Chapter XI). Let $G' + B'$ be a DIRICHLET region, and $G'_n + B'_n$ a sequence approximating G' in the sense of 5. 6. Denote by F a function which is continuous on $G' + B'$, by h the solution of the DIRICHLET problem for $G' + B'$ with the boundary condition $h = F$ on B', and by h_n the solution of the DIRICHLET problem for G'_n with the boundary condition $h_n = F$ on B'_n. Then h_n approximates h in the following sense. Given $\varepsilon > 0$, we have an $n_0 = n_0(\varepsilon)$ such that $|h - h_n| < \varepsilon$ in $G'_n + B'_n$ for $n > n_0$. This follows by simple ε-arguments from the maximum-minimum principle for harmonic functions.

5. 24. *Remarks on the formula of* GREEN. Let g be continuous together with its derivatives of the first and second order in a domain G. Consider a region $G' + B'$ comprised in G, such that B' consists of a finite number of non-intersecting smooth JORDAN curves. Take a point (x_0, y_0) in G', and take r small enough so that the closed circular disc with centre (x_0, y_0) and radius r is comprised in G'. Put (see 4. 19 to 4. 21 for notations)

$$l(x, y) = l(x, y; x_0, y_0)\,, \qquad l_r(x, y) = l_r(x, y; x_0, y_0)\,,$$

$$g^{(r)}(x_0, y_0) = \frac{1}{2\pi}\int\limits_0^{2\pi} g(x_0 + r\cos\varphi,\ y_0 + r\sin\varphi)\, d\varphi\,.$$

Denote by H the solution of the DIRICHLET problem for G' with the boundary condition $H = l$ on B', and by h the solution of the DIRICHLET problem for G' with the boundary condition $h = g$ on B'. The functions H, l, l_r depend also upon (x_0, y_0), but this point will be kept fixed and therefore it is unnecessary to use notations like $H(x, y; x_0, y_0)$. The function h is independent of (x_0, y_0), and H, h, l are all independent of r. As g and B' are smooth, it is easy to justify the application of GREEN's identity (KELLOGG [1], p. 215) in deriving the formula

$$g^{(r)}(x_0, y_0) = -\frac{1}{2\pi} \iint\limits_{G'} [l_r(x, y) - H(x, y)] \, \Delta g(x, y) \, dx \, dy + h(x_0, y_0) \, .$$

5.25. For $r \to 0$ we obtain the classical formula

$$g(x_0, y_0) = -\frac{1}{2\pi} \iint\limits_{G'} \mathfrak{G}(x, y; x_0, y_0) \, \Delta g(x, y) \, dx \, dy + h(x_0, y_0) \, ,$$

where $\mathfrak{G} = l - H$ is GREEN's function for G' with pole at (x_0, y_0). Conversely, an integration leads back to the formula of 5.24 which is more convenient in some applications.

5.26. Drop now the assumption that the boundary B' of G' is smooth and suppose only that $G' + B'$ is a DIRICHLET region. Otherwise, let all assumptions and notations stand as in 5.24. *Then the formula of 5.24 still holds.* This is easily proved, on the basis of 5.23, by applying the formula to a sequence of regions which approximate G' in the sense of 5.6.

5.27. If the function g of 5.26 is subharmonic in G, then the function h is the B. H. M. of g in G' (observe that g is continuous by assumption and use 5.4).

5.28. We proceed to discuss the question raised in 5.8. We start with the following theorem. Let u be subharmonic in a domain G. Denote by $G' + B'$ a DIRICHLET region comprised in G, and by $\bar{h}$ the B. H. M. of u in G'. Suppose that u is *harmonic* in G'. Then $\bar{h} = u$ in G' (RADÓ [4]).

5.29. To prove this theorem, consider the approximating functions $u_k^{(3)}$ defined in 2.21. If $G'' + B''$ is a region such that $G' + B' \subset G''$, $G'' + B'' \subset G$, then for large k the function $u_k^{(3)}$ is defined in G'' and is continuous there together with its derivatives of the first and second order. Also, $u_k^{(3)}$ is subharmonic, and hence $\Delta u_k^{(3)} \geq 0$. By 2.24 we have

$$0 \leq \iint\limits_{G'} \Delta u_k^{(3)}(x, y) \, dx \, dy < N$$

where N is a finite constant. If S is any closed set in G', then we have $u_k^{(3)} = u$ and $\Delta u_k^{(3)} = 0$ on S for large k (see 2.23). Take now any point (x_0, y_0) in G' and a small r. For large k, the function $u_k^{(3)}$

is then harmonic on the closed circular disc with centre (x_0, y_0) and radius r (see 2. 23) and hence we have

$$\frac{1}{2\pi} \int_0^{2\pi} u_k^{(3)} (x_0 + r \cos\varphi, y_0 + r \sin\varphi)\, d\varphi = u_k^{(3)} (x_0, y_0) = u(x_0, y_0).$$

5. 30. Using the preceding facts, we obtain from 5. 26 the formula

$$u_k^{(3)} (x_0, y_0) = -\frac{1}{2\pi} \iint_{G'} [l_r(x, y) - H(x, y)]\, \Delta u_k^{(3)} (x, y)\, dx\, dy + h_k^{(3)} (x_0, y_0),$$

where $h_k^{(3)}$ is the solution of the DIRICHLET problem for G' with the boundary condition $h_k^{(3)} = u_k^{(3)}$ on B'.

5. 31. As $u_k^{(3)}$ is continuous and $u_k^{(3)} \searrow u$ on B' (see 2. 21), we have $h_k^{(3)} \searrow \bar{h}$ for $k \to \infty$, where $\bar{h}$ is the B. H. M. of u in G' (see 5. 4).

5. 32. Give now an $\varepsilon > 0$. Observe that $l_r - H$ is continuous on $G' + B'$ and $l_r - H = l - H = 0$ on B'. Hence we have a $\delta > 0$ such that $|l_r - H| < \varepsilon$ in $G' - S_\delta$, where S_δ denotes the (closed) set of all those points in G' whose distance from B' is $\geq \delta$. We write now

$$\iint_{G'} [l_r(x, y) - H(x, y)]\, \Delta u_k^{(3)} (x, y)\, dx\, dy = \iint_{S_\delta} + \iint_{G' - S_\delta} = I_k^{(1)} + I_k^{(2)}.$$

By 5. 29 we have then $I_k^{(1)} = 0$ and $|I_k^{(2)}| < \varepsilon N$ for large k. As ε is arbitrary, it follows that the integral in the formula of 5. 30 converges to zero. The term $h_k^{(3)} (x_0, y_0)$ in that formula converges to $\bar{h}(x_0, y_0)$ (see 5. 31). Thus the theorem of 5. 28 follows from the formula of 5. 30 for $k \to \infty$.

5. 33. Denote by $G' + B'$ a region comprised in the domain G in which u is subharmonic. Consider a function h' which is harmonic in G' and define in G a function u' as follows: $u' = u$ in $G - G'$ and $u' = h$ in G'. If u' is subharmonic in G, then let us say that h' is *admissible* for u in G'. We have then the theorem: if $G' + B'$ is a DIRICHLET region comprised in G, then there exists in G' *exactly one* harmonic function which is admissible for u in G' (RADÓ [4]). The fact that there exists *at most one* admissible harmonic function follows immediately from 5. 28 and 5. 4. The fact that there exists *at least one* follows from the next theorem.

5. 34. If $G' + B'$ is a DIRICHLET region comprised in G, then the best harmonic majorant $\bar{h}$ of u in G' (see 5. 4) is admissible for u in G' (EVANS [4], part I, p. 237). This follows immediately from the definition of the best harmonic majorant.

5. 35. If $G' + B'$ is a DIRICHLET region comprised in G and if $\bar{h}$ and h^* denote the B. H. M. and the L. H. M. of u in G', then $\bar{h} \equiv h^*$ (RADÓ [4]). On account of 5. 33 and 5. 34 this will be proved if we show that h^* is admissible for u in G', and this fact follows immediately from 5. 34 and from the relation $u \leq h^* \leq \bar{h}$.

Chapter VI.

Representation of subharmonic functions in terms of potentials.

6.1. Every sufficiently smooth function v can be represented as a potential plus a harmonic function (KELLOGG [1], p. 219). We shall formulate this fact in a form suitable for our purposes. Let v be continuous in a domain G together with its derivatives of the first and second order. Take a region $G' + B'$ in G, such that B' consists of a finite number of non-intersecting smooth JORDAN curves. From GREENS identity we obtain the formula

$$L(v; P; r) = \frac{1}{2\pi} \iint\limits_{G'} l_r(P, Q)\, \Delta v(Q)\, da_Q + h(P), \quad P \text{ in } G', \ r \text{ small,}$$

where

$$h(P) = \frac{1}{2\pi} \int\limits_{B'} \left(u(Q) \frac{\partial l(P, Q)}{\partial n_e} - l(P, Q) \frac{\partial u(Q)}{\partial n_e} \right) ds$$

is harmonic in G' (see 1.5, 4.19, 4.21, 4.17 for notations).

6.2. The harmonic function h of 6.1 depends only upon the values of v on B' and in the vicinity of B' (see the explicit formula in 6.1). In particular, h is independent of r. For $r \to 0$ we obtain (cf. 4.33)

$$v(P) = \int\limits_{G'} l(P, Q)\, d\mu(e_Q) + h(P), \quad P \text{ in } G',$$

where μ is the mass-distribution with density $-(1/2\pi)\, \Delta v$.

6.3. It is a fundamental result of F. RIESZ that every subharmonic function admits of a representation of this form, regardless of its possible lack of smoothness. F. RIESZ ([5], part II) gave two proofs for this theorem. We shall sketch a simplified version, due to G. C. EVANS ([4], part I, p. 237), of the second proof of RIESZ.

6.4. *A selection theorem* (special case of RADON [1], p. 1337; see also RIESZ [5], part II, p. 351). Let there be given on a closed set S a sequence of positive mass-distributions $\mu_k(e)$ such that $\mu_k(S)$ is less than some finite constant independent of k. Then there exists a subsequence μ_{k_r} and a positive mass-distribution $\mu(e)$ on S, such that

$$\int\limits_S f(Q)\, d\mu_{k_r}(e_Q) \to \int\limits_S f(Q)\, d\mu(e_Q)$$

for every function $f(Q)$ which is continuous on S. The subsequence μ_{k_r} is then said to converge *weakly* to μ on S.

6.5. Consider now a function $u(x, y) = u(P)$ which is subharmonic in a domain G. Take a region $G' + B'$ comprised in G. Take an auxiliary region $G'' + B''$ such that $G' + B' \subset G''$, $G'' + B'' \subset G$ and B''

consists of a finite number of smooth non-intersecting JORDAN curves Γ_i'', $i = 1, 2, \ldots, j$. Denote by D_i'' a narrow doubly connected DIRICHLET domain which contains Γ_i'' in its interior and by $\bar{h}_i$ the B. H. M. of u in D_i''. Define a function $\bar{u}$ in G as follows: $\bar{u} = u$ in $G - \Sigma D_i''$ and $\bar{u} = \bar{h}_i$ in D_i'', $i = 1, 2, \ldots, j$. Then $\bar{u}$ is subharmonic in G (see 5.34). Also, $\bar{u}$ is harmonic on B'' and in the vicinity of B'', and $\bar{u} = u$ in and near to $G' + B'$.

6. 6. Consider now the functions $\bar{u}_k^{(3)}(x, y) = A_{1/k}(x, y; \bar{u})$ defined in 2. 21. We have by 6. 1 the formula

$$L(\bar{u}_k^{(3)}; P; r) = \frac{1}{2\pi} \iint\limits_{G''} l_r(P, Q)\, \Delta \bar{u}_k^{(3)}(Q)\, da_Q + h_k(P),$$

$$P \text{ in } G'', \; k \text{ large}, \; r \text{ small}.$$

6. 7. By 2. 23 we have $\bar{u}_k^{(3)} = \bar{u}$ on B'' and in the vicinity of B''. *Hence h_k is independent of k*, because h_k depends only upon the values of $\bar{u}_k^{(3)}$ on and near to B'' (see 6. 2). We can write therefore h instead of h_k.

6. 8. Define, for large k,

$$\bar{\mu}_k(e) = \frac{1}{2\pi} \iint\limits_{e} \Delta \bar{u}_k^{(3)}(Q)\, da_Q, \quad e \subset G'' + B''.$$

We have then by GREEN's identity

$$\bar{\mu}_k(G'' + B'') = \frac{1}{2\pi} \int\limits_{B''} \frac{\partial \bar{u}_k^{(3)}}{\partial n_e}\, ds = \frac{1}{2\pi} \int\limits_{B''} \frac{\partial \bar{u}}{\partial n_e}\, ds.$$

Hence we can apply the selection theorem of 6. 4 and we obtain on $G'' + B''$ a positive mass-distribution $\mu(e)$, such that a certain subsequence $\bar{\mu}_{k_\nu}$ converges weakly to μ on $G'' + B''$.

6. 9. By 6. 7 and 4. 17 the formula of 6. 6 can be written in the form

$$L(\bar{u}_k^{(3)}; P; r) = \int\limits_{G'' + B''} l_r(P, Q)\, d\mu_k(e_Q) + h(P), \quad P \text{ in } G'',$$

since $\mu_k(B'')$ is clearly equal to zero. For $k = k_\nu$, $\nu \to \infty$ we obtain by 6. 8 and 2. 21

$$L(\bar{u}; P; r) = \int\limits_{G'' + B''} l_r(P, Q)\, d\mu(e_Q) + h(P), \quad P \text{ in } G''.$$

For P in G' we have $\bar{u} = u$ by 6. 5, and for $r \to \infty$ it follows by 2. 7 and 4. 23 that

$$u(P) = \int\limits_{G'' + B''} l(P, Q)\, d\mu(e_Q) + h(P) = \int\limits_{G'} + \int\limits_{G'' + B'' - G'} + h(P)$$

for P in G'. The second integral on the right is a harmonic function of P in G' (see 4. 25). Hence we have the following theorem.

6.10. If u is subharmonic in a domain G, and if G' is a domain completely interior to G, then there exists in G' a positive mass-distribution $\mu(e)$ such that

$$u(P) = -\int_{G'} \log \frac{1}{PQ} \, d\mu(e_Q) + H(P), \quad P \text{ in } G',$$

where H is harmonic in G' (RIESZ [5], part II).

6.11. We shall see now that the distribution $\mu(e)$ is *unique*. Put

$$v(P) = -\int_{G'} \log \frac{1}{PQ} \, d\mu(e_Q), \quad P \text{ in } G'.$$

Take in G' any two JORDAN curves C_1, C_2 such that C_1 is enclosed by C_2 and the interior of C_2 is comprised in G'. We have then (see 5.11) $F(C_1, C_2; u) = F(C_1, C_2; v) + F(C_1, C_2; H)$. But $F(C_1, C_2; H) = 0$, since H is harmonic in and on C_2. Hence $F(C_1, C_2; v)$ is univocally determined by u, and by 5.20 and 4.14 it follows that $\mu(e)$ is univocally determined on G'.

6.12. More generally, consider two domains G_1', G_2' completely interior to G, and denote by $\mu_1(e), \mu_2(e)$ the distributions which correspond to u_1, u_2 in the sense of 6.10 and 6.11. Then $\mu_1(e) = \mu_2(e)$ for every set e of class (B) which is comprised in $G_1' G_2'$. This follows by a reasoning similar to that in 6.11.

6.13. Let G be a domain and $\overset{\circ}{\mu}(e)$ a set-function which is defined only for sets e which are *completely interior* to G (that is, the limit points of e are also comprised in G; we only consider sets e which are measurable in the BOREL sense). If otherwise $\overset{\circ}{\mu}(e)$ possesses the properties required in 4.7, then $\overset{\circ}{\mu}(e)$ will be called a *generalized* positive mass-distribution on G. For such a distribution $\overset{\circ}{\mu}$ it might happen that the least upper bound of $\overset{\circ}{\mu}(e)$, for all sets e completely interior to G, is equal to $-\infty$.

6.14. Let E be a set [measurable (B)] completely interior to G. Considered on E, the $\overset{\circ}{\mu}$ of 6.13 is clearly a positive mass-distribution in the original sense of 4.7. Hence we can consider on E STIELTJES-RADON integrals in terms of $\overset{\circ}{\mu}$.

6.15. If u is subharmonic in a domain G, then there exists on G a univocally determined generalized positive mass-distribution $\overset{\circ}{\mu}(e)$ (see 6.13), such that for every domain G' completely interior to G we have

$$u(P) = -\int_{G'} \log \frac{1}{PQ} \, d\overset{\circ}{\mu}(e_Q) + h(P), \quad P \text{ in } G',$$

where h is harmonic in G' (RIESZ [5], part II; cf. 5.22). This follows immediately from 6.10 and 6.12.

6.16. Consider a DIRICHLET region $G' + B'$ and a point P in G'. Then GREEN's function for G' with pole at P is defined by $\mathfrak{G}(P, Q) = \log(1/PQ) - H(P, Q)$ where $H(P, Q)$ is the solution of the DIRICHLET problem for G' with the boundary condition $H(P, Q) = \log(1/PQ)$ on B'. Consider next a general bounded domain G. Approximate G by a sequence $G_n + B_n$ of DIRICHLET regions as explained in 5.6. If P is a point in G, then P will be in G_n for large n. GREEN's function $\mathfrak{G}(P, Q)$ for G with pole at P is then defined by $\mathfrak{G}(P, Q) = \lim \mathfrak{G}_n(P, Q)$. $\mathfrak{G}(P, Q)$ is a finite, positive and harmonic function of Q in G, except for $Q = P$, and we have $\mathfrak{G}(P, Q) = \log(1/PQ) - H(P, Q)$, where $H(P, Q)$ is a harmonic function of Q in G, even for $Q = P$ (see KELLOGG [1], Chapter IX for information concerning GREEN's function).

6. 17. Let there be given in a bounded domain G a positive mass-distribution $\mathring{\mu}(e)$ in the generalized sense of 6. 13. Consider

$$v_k(P) = -\int\limits_{G_k} \mathfrak{G}(P, Q)\, d\mathring{\mu}(e_Q), \qquad P \text{ in } G_k,$$

where the sequence G_k approximates G in the sense of 5.6. We have more explicitly (see 6. 16)

$$v_k(\dot{P}) = -\int\limits_{G_k} \log \frac{1}{PQ}\, d\mathring{\mu}(e_Q) + \int\limits_{G_k} H(P, Q)\, d\mathring{\mu}(e_Q).$$

Thus v_k appears as the sum of a subharmonic function and of a harmonic function. Hence v_k is subharmonic in G_k. Clearly v_k decreases if k increases. By 3. 6, either $v_k \to -\infty$ everywhere in G, or v_k converges to a subharmonic function v in G. In the latter case we write

$$v(P) = -\int\limits_{G} \mathfrak{G}(P, Q)\, d\mathring{\mu}(e_Q).$$

On account of its definition, this integral is therefore a subharmonic function of P whenever it exists. The value of the integral is easily seen to be independent of the sequence G_k.

6. 18. Consider now a function u which is subharmonic in the bounded domain G. Denote by $\mathring{\mu}$ the generalized distribution which corresponds to u in the sense of 6. 15. Then the integral of 6. 17 exists if and only if we have some harmonic function which is $\geq u$ in G. If this condition is satisfied then

$$\int\limits_{G} \mathfrak{G}(P, Q)\, d\mathring{\mu}(e_Q) \leq h(P) - u(P)$$

for every harmonic function which is $\geq u$ in G (RIESZ [5], part II). Proof. Suppose first that we have a harmonic function $h \geq u$ in G. Take a domain G' completely interior to G and introduce again the auxiliary region $G'' + B''$ and the auxiliary functions $\bar{u}$, $\bar{u}_k^{(3)}$ as in 6. 5 and 6. 6. Denote by $\mathfrak{G}''_r(P, Q)$ the function obtained from GREEN's function for G'' if we replace $l(P, Q)$ by $l_r(P, Q)$ (see 4. 21). We have then by 5. 24, 6. 7, 6. 8 the formula

$$L(\bar{u}_k^{(3)}; P; r) = -\int\limits_{G'' + B''} \mathfrak{G}''_r(P, Q)\, d\mu_k(e_Q) + \bar{h}_k(P), \qquad P \text{ in } G'', \ k \text{ large}, \ r \text{ small},$$

where $\bar{h}_k$ is the solution of the DIRICHLET problem for G'' with the boundary condition $\bar{h}_k = \bar{u}_k^{(3)}$ on B''. For P in G' it follows, by a reasoning similar to that in 6.5 to 6.9, that

$$L(\bar{u};P;r) = -\int_{G''+B''} \mathfrak{G}''_r(P,Q)\,d\mu(e_Q) + \bar{h}(P)\,, \qquad P \text{ in } G',$$

where the harmonic function $\bar{h}$ is determined by the condition $\bar{h} = \bar{u}$ on B''. It follows from the definition of $\bar{u}$ (see 6.5) that $\bar{h} \leq h$ in $G'' + B''$ and $u \leq \bar{u}$ in G. We have therefore

$$h(P) - L(u;P;r) \geq \bar{h}(P) - L(\bar{u};P;r) \geq \int_{G'} \mathfrak{G}''_r(P,Q)\,d\overset{\circ}{\mu}(e_Q)\,,$$

since $\overset{\circ}{\mu}(e) = \mu(e)$ on G', by 6.8, 6.9, 6.10, 6.15. Denote by $\mathfrak{G}_r(P,Q)$ the function obtained from GREEN's function for G if we replace $l(P,Q)$ by $l_r(P,Q)$ (see 4.21). Let G'' approach G in the sense of 5.6. Then $\mathfrak{G}''_r \nearrow \mathfrak{G}_r$ and by a well-known theorem ot DINI the convergence is uniform on every closed set in G (and hence on every set completely interior to G), since $\mathfrak{G}''_r$ and $\mathfrak{G}_r$ are continuous. We obtain for $G'' \to G$ the inequality $h(P) - L(u;P;r) \geq \int_{G'} \mathfrak{G}_r(P,Q)\,d\overset{\circ}{\mu}(e_Q)$. For $r \to 0$ it follows, by 6.17, that $h(P) - u(P) \geq \int_{G'} \mathfrak{G}(P,Q)\,d\overset{\circ}{\mu}(e_Q)$ for P in G'.

As G' is any domain completely interior to G, the preceding inequality proves both the existence of the integral $\int_{G} \mathfrak{G}(P,Q)\,d\overset{\circ}{\mu}(e_Q)$ and the inequality asserted in the theorem. Conversely, suppose that the preceding integral exists. Consider the functions v_k of 6.17 relative to the distribution $\overset{\circ}{\mu}$ which corresponds to the given subharmonic function u. By the definition of $\overset{\circ}{\mu}$ (see 6.15) we have $u(P) = v_k(P) + h_k(P)$ for P in G_k, where h_k is harmonic in G_k. If k increases, v_k decreases and hence h_k increases. By the theorem of HARNACK, $h_k \to h^*$ where either h^* is $\equiv +\infty$ in G or h^* is harmonic in G. Clearly the first case is incompatible with our present assumptions. For $k \to \infty$ we obtain therefore (cf. 6.17)

$$u(P) = -\int_{G} \mathfrak{G}(P,Q)\,d\overset{\circ}{\mu}(e_Q) + h^*(P)\,, \qquad P \text{ in } G.$$

But $\overset{\circ}{\mu}$ and $\mathfrak{G}$ are both positive, and hence $h^* \geq u$ in G. The existence of a harmonic majorant for u in G is proved.

6.19. The harmonic function h^* of the last formula of 6.18 is actually the least harmonic majorant of u in G. Indeed, if h is any harmonic majorant of u in G, then we have by 6.18

$$h^*(P) = u(P) + \int_{G} \mathfrak{G}(P,Q)\,d\overset{\circ}{\mu}(e_Q) \leq u(P) + (h(P) - u(P)) = h(P)\,.$$

We have therefore the theorem: If u is subharmonic in a bounded domain G and if there exists a harmonic majorant for u in G, then u can be represented in the form

$$u(P) = -\int_G \mathfrak{G}(P, Q)\, d\mathring{\mu}(e_Q) + h^*(P), \quad P \text{ in } G,$$

where h^* is the least harmonic majorant of u in G and $\mathring{\mu}$ is a generalized positive mass-distribution on G (Riesz [5], part II).

6. 20. If u is a *smooth* subharmonic function, then the corresponding distribution can be expressed in terms of the Laplacian Δu (see 6. 2 and 6. 11). It is then natural to expect that the preceding theorems can be discussed in terms of the generalized Laplacians introduced by various authors. It seems that no explicit discussion was given as yet on this basis (cf. the remarks of F. Riesz in Wiener [1], p. 7).

6. 21. In the light of the theorem of 6. 10, the theory of subharmonic functions appears as a chapter in potential theory. It is beyond the scope of this report to follow up the implications of this situation. The reader desiring further information will find a wealth of interesting material and a large number of references in Frostman [1], Evans [4], Kellogg [1].

6. 22. The theorem of 6. 10 implies that the study of subharmonic functions *in the small* can be based upon a study of the potential $v(P) = \int \log PQ\, d\mu(e_Q)$. In the way of illustration, we mention two results obtained in this manner. According to Evans ([4], part I, pp. 233—235) the potential $v(P)$ is an absolutely continuous function of x for almost every y and an absolutely continuous function of y for almost every x. As a consequence, the partial derivatives v_x and v_y exist almost everywhere. It follows by further discussion that v_x and v_y are summable on every bounded measurable set. The application to subharmonic functions is immediate on account of 6. 10.

6. 23. Using the notations of 6. 15, consider the integral mean $A_r(x, y; u)$ (see 2. 19). As $A_r(x, y; u)$ is again subharmonic, it will give rise to a distribution $\mathring{\mu}_r(e)$ in the sense of 6. 15. It might be expected that $\mathring{\mu}_r(e)$ will be smoother than the distribution $\mathring{\mu}(e)$ corresponding to u itself. By means of 6. 15 it follows from results of Thompson [1] that $\mathring{\mu}_r(e)$ is a distribution with a summable density $\delta_r(P)$ given by

$$\delta_r(P) = \frac{1}{r^2 \pi}\, \mathring{\mu}(C(r; P)).$$

where $C(r; P)$ denotes the interior of the circle with centre P and radius r. The proof depends upon a discussion of changes in the order of integrations in iterated Stieltjes-Radon integrals.

Chapter VII.

Analogies between harmonic and subharmonic functions.

7.1. The general theory of subharmonic functions, as sketched in the preceding Chapters, was based largely upon a few elementary properties of harmonic functions. Practically every paper quoted in this report contains interesting developments concerned with the implications of more involved properties of harmonic functions. The purpose of this Chapter is to give a picture of the results obtained in this direction. The reader will note that the proofs sketched in the sequel do not always apply in the case of more than two variables. Such situations lead to interesting problems, some of which seem to be quite difficult. As a first topic, we shall consider *isolated singularities of subharmonic functions*. If u is known to be subharmonic in the vicinity of a point (x_0, y_0), this point itself being excluded, then (x_0, y_0) will be called an isolated singular point of u. Without loss of generality we can assume that (x_0, y_0) is the point $O = (0, 0)$. We shall review presently some results of BRELOT [1]. Various details of the following presentation are based on unpublished remarks of S. SAKS.

7.2. (See 1.5 for notations.) Put $L(u; r) = L(u; 0, 0; r)$, $\lambda = u/\log(1/r)$, $L(\lambda; r) = L(u; r)/\log(1/r)$. By 2.5, $L(u; r)$ is a convex function of $\log r$ and hence of $\log(1/r)$ for small r. Using some simple properties of convex functions, we obtain a number of facts concerning $L(u; r)$ and $L(\lambda; r)$ (BRELOT [1], pp. 23—37), some of which we shall list now explicitly.

7.3. For $r \to 0$ both $L(u; r)$ and $L(\lambda; r)$ converge to definite (not necessarily finite) limits which will be denoted by $L(u; 0)$ and $L(\lambda; 0)$ respectively. For small values of r both $L(u; r)$ and $L(\lambda; r)$ are *monotonic*, and for $r \searrow 0$ either $L(u; r) \nearrow L(u; 0) = +\infty$ or $L(u; r) \searrow L(u; 0) \geq -\infty$, and either $L(\lambda; r) \nearrow L(\lambda; 0) \leq +\infty$ or $L(\lambda; r) \searrow L(\lambda; 0) > -\infty$. Note that if $L(u; r)$ increases for $r \searrow 0$ then always $L(u; 0) = +\infty$, and if $L(\lambda; r)$ decreases for $r \searrow 0$ then always $L(\lambda; 0) > -\infty$.

7.4. We shall use $\overset{+}{a}$ to denote the greater one of the numbers a and zero. Clearly $a \leq \overset{+}{a} \leq |a|$ and $|a| = 2\overset{+}{a} - a$.

7.5. Let us recall a few facts concerning isolated singularities of *harmonic* functions. If $h(P)$ is harmonic in the vicinity of O with the possible exception of O itself, then we have the expansion (see KELLOGG [1], Chapter XII)

$$h(P) = h_0(P) + \gamma \log \frac{1}{OP} + \sum_{n=1}^{\infty} \frac{\alpha_n \cos n\varphi + \beta_n \sin n\varphi}{OP^n} = h_0(P) + h_1(P),$$

where $h_0(P)$ is harmonic even at O, $h_1(P)$ is harmonic in the whole plane with the possible exception of O, φ is the polar angle determined by $x = OP\cos\varphi$, $y = OP\sin\varphi$, and γ, α_n, β_n are constants. Suppose that $h \geq 0$ in the vicinity of O. Then $\alpha_n = \beta_n = 0$ for $n = 1, 2, \ldots$ (see BRELOT [1] for references; see also RIESZ [5], part II, p. 350). Indeed, we have $L(h\cos n\varphi; r) = h_0(O) + \gamma\log(1/r) + \alpha_n/(2r^n)$. Hence $\alpha_n = 2\lim r^n L(h\cos n\varphi; r)$, $r \to 0$. But $|L(h\cos n\varphi; r)| \leq L(|h|; r) = L(h; r) = h_0(O) + \gamma\log(1/r)$. Thus $r^n L(h\cos n\varphi; r) \to 0$ for $r \to 0$, and hence $\alpha_n = 0$. The coefficient β_n is discussed in the same way.

7.6. Until further notice, u denotes a subharmonic function which has an isolated singularity at O. The following remark will be useful in the sequel. Consider, for small r'', a circular ring $R: 0 < r' < (x^2+y^2)^{1/2} < r''$, and denote by $\bar{h}$ the B.H.M. of u in R (see 5.4). Then (cf. 1.12, 1.13) $L(\bar{h}; r) = L(\bar{h}; 0, 0; r)$ is a linear function $a\log(1/r) + b$ of $\log(1/r)$ and we have $a\log(1/r') + b = L(u; r')$, $a\log(1/r'') + b = L(u; r'')$. Suppose we are given inequalities $L(u; r') \leq A\log(1/r') + B$, $L(u; r'') \leq A\log(1/r'') + B$, where A, B are constants. Clearly, it follows that $L(\bar{h}; r) \leq A\log(1/r) + B$ for $r' < r < r''$.

7.7. Since u is subharmonic in the vicinity of O, the function $\overset{+}{u} = \overline{u, 0}$ is also subharmonic there (see 3.4). We shall use the symbols $L(\overset{+}{u}; r)$, $L(\overset{+}{\lambda}; r)$ in the sense of 7.2.

7.8. We shall use $V - O$ to denote a vicinity of O, less O itself. It is assumed that u is subharmonic in $V - O$ and on the boundary of $V - O$, except for the point O. Vicinities of the form $0 < x^2 + y^2 < \varrho^2$ will be denoted by $V_\varrho - O$. It will be understood that ϱ is so small that $L(u; r)$, $L(\lambda; r)$, $L(\overset{+}{u}; r)$, $L(\overset{+}{\lambda}; r)$ are *monotonic* for $0 < r < \varrho$ (see 7.3 and 7.7).

7.9. Suppose that in a vicinity $V_\varrho - O$ we have a harmonic majorant H for u, and let $H = H_0 + H_1$ be the expansion of H (cf. 7.5). Let $V - O$ be any vicinity which contains $V_\varrho - O$. Then u has in $V - O$ a harmonic majorant of the form $H_1 + $ const. This follows immediately from the fact that u is upper semi-continuous and H_1 is continuous in $V - O$ and on the boundary of $V - O$, the point O being excluded.

7.10. Generally there will not exist a harmonic majorant for u in the vicinity of O. However, if we have a harmonic majorant $H^{(1)}$ in a vicinity $V_1 - O$, then we also have a harmonic majorant in any other vicinity $V_2 - O$ (BRELOT [1], p. 32). To see this, take a vicinity $V_\varrho - O$ comprised both in $V_1 - O$ and in $V_2 - O$. Observe that $H^{(1)}$ is also a majorant in $V_\varrho - O$ and apply 7.9.

7.11. Suppose u has a harmonic majorant in a vicinity $V_\varrho - O$. Then (see 5.7) we have in $V_\varrho - O$ a L.H.M. h^*. Let $h^* = h_0 + h_1$ be the expansion of h^* (see 7.5). Consider now any vicinity $\tilde{V} - O$

containing $V_\varrho - O$. By 7.9 we have in $\tilde{V} - O$ a harmonic majorant $\tilde{H}$ of the form $h_1 +$ const., and hence the L. H. M. $\tilde{h}^*$ of u in $\tilde{V} - O$ satisfies an inequality $\tilde{h}^* < h_1 +$ const. in the vicinity of O. We have therefore $0 \le h^* - \tilde{h}^* \le h_0 +$ const. in the vicinity of O. Consequently (cf. 7.5) $h^* - \tilde{h}^*$ is harmonic even at O.

7. 12. If h_1^*, h_2^* are the L. H. M. of u in the vicinities $V_1 - O$, $V_2 - O$ respectively, then $h_1^* - h_2^*$ is harmonic even at O (Brelot [1], p. 32). Proof. Take a vicinity $V_\varrho - O$ comprised both in $V_1 - O$ and in $V_2 - O$ and apply 7.11.

7. 13. Suppose that u has a harmonic majorant in a vicinity of O. By 7.6 we have then a L. H. M. for u in every vicinity $V - O$ and by 7.11, 7.12, 7.5 the constants γ, α_n, β_n have the same values in the expansions of all these least harmonic majorants.

7. 14. u has a harmonic majorant in the vicinity of O if and only if $L(\lambda; 0)$ (see 7.3) is finite. If this condition is satisfied, then the constant γ of 7.13 is equal to $L(\lambda; 0)$ (Brelot [1], p. 32). Proof. The necessity of the condition is obvious. To prove the sufficiency, assume that $L(\lambda; 0)$ is finite. Give an $\varepsilon > 0$ and take a small r_0 such that $L(\lambda; r) < L(\lambda; 0) + \varepsilon$ for $0 < r < r_0$. Take any r such that $0 < r < r_0$ and take a sequence r_n such that $r > r_1 > r_2 > \cdots \to 0$. Denote by $\bar{h}_n$ the B. H. M. of u in the ring $r_n < (x^2 + y^2)^{1/2} < r_0$. We have then $L(u; r_n) < (L(\lambda; 0) + \varepsilon) \log(1/r_n)$, $L(u; r_0) < (L(\lambda; 0) + \varepsilon) \log(1/r_0)$, and hence, by 7.6, $L(\bar{h}^n; r) < (L(\lambda; 0) + \varepsilon) \log(1/r)$. This shows that $\bar{h}_n$ cannot converge to $+\infty$ everywhere. By 5.5 we have $\bar{h}_{n+1} \ge \bar{h}_n \ge u$ in the ring $r_n < (x^2 + y^2)^{1/2} < r_0$. Hence, by the theorem of Harnack, $\bar{h}_n$ converges in $V_{r_0} - O$ to a harmonic function $h^* \ge u$, and the existence of a harmonic majorant is proved. By 5.7, h^* is the L. H. M. of u in $V_{r_0} - O$. Also, the inequality $L(\bar{h}_n; r) < (L(\lambda; 0) + \varepsilon) \log(1/r)$ implies that we have $L(h^*; r) \le (L(\lambda; 0) + \varepsilon) \log(1/r)$ for $0 < r < r_0$. To estimate the constant γ (cf. 7.5, 7.13) in the expansion of h^*, observe that

$$\gamma = \lim_{r \to 0} \frac{L(h^*; r)}{\log \dfrac{1}{r}} \le L(\lambda; 0) + \varepsilon.$$

On the other hand, $u \le h^*$ in $V_{r_0} - O$ and hence

$$L(\lambda; 0) = \lim_{r \to 0} \frac{{}^!L(u; r)}{\log \dfrac{1}{r}} \le \lim_{r \to 0} \frac{L(h^*; r)}{\log \dfrac{1}{r}} = \gamma.$$

As ε is arbitrary, it follows that $\gamma = L(\lambda; 0)$.

7. 15. If u is bounded from above in the vicinity of O, then u is subharmonic even at O (Brelot [1], p. 27). Instead of reproducing the proof of Brelot, let us observe that this follows immediately from the theorems of 3.35 and 3.37 on almost subharmonic functions. Indeed, consider for $n = 1, 2, \ldots$ the function $u_n(P) = u(P) - (1/n) \log(1/OP)$ for $P \neq O$, $u_n(O) = -\infty$. Clearly, u_n is subharmonic in a small disc $D: x^2 + y^2 < r^2$, even at O, since by assumption $u < M$ in $D - O$, where M is some finite constant. We have $u_n \leq u_{n+1} < M$ in D, since $u_{n+1}(O) = -\infty$. Hence, by 3.37, the limit function $u^* = \lim u_n$ is almost subharmonic in D. We have therefore in D a subharmonic function $\bar{u}$ such that $\bar{u} = u^*$ almost everywhere in D. But $u^* = u$ in $D - O$, and hence $\bar{u} = u$ in $D - O$ by 3.35. As $\bar{u}$ is subharmonic even at O, the theorem is proved.

7.16. If u has a harmonic majorant H in the vicinity of O, then $v = u - H$ is subharmonic even at O (BRELOT [1], p. 35). As $v \leq 0$ in the vicinity of O, this follows immediately from 7.15.

7.17. (See 7.2, 7.3, 7.7 for notations). If $L(\overset{+}{u}; 0)$ is finite, then u remains subharmonic at O (BRELOT [1], pp. 34—35). More generally, if $L(\overset{+}{\lambda}; 0) = 0$, then u remains subharmonic at O (SAKS, unpublished). Proof. By 7.7 and 7.14, $\overset{+}{u}$ has a harmonic majorant in the vicinity of O. Denote by H^* the least harmonic majorant of $\overset{+}{u}$ in a vicinity $V_\varrho - O$. Then $H^* \geq \overset{+}{u} \geq 0$ and hence by 7.5 we have $H^*(P) = H_0(P) + \Gamma \log(1/OP)$, where H_0 is harmonic even at O. By 7.14 we have $\Gamma = L(\overset{+}{\lambda}; 0)$, which is equal to zero by assumption. Hence H^* is harmonic even at O. Consequently $\overset{+}{u}$ is bounded from above in the vicinity of O. As $u \leq \overset{+}{u}$, the theorem follows now by 7.15.

7.18. The work of BRELOT contains a number of further results and applications which cannot be reproduced here. We shall review presently certain results concerned with generalizations of properties of harmonic functions in the vicinity of the boundary of the domain of definition. The first results in this direction were obtained by LITTLEWOOD [2, 3, 4] and EVANS [2]. These results were later on extended by EVANS [3] and PRIVALOFF [1, 2]. EVANS obtained his results by methods in Potential Theory. PRIVALOFF extended some results obtained by LITTLEWOOD in the special case of the circle to more general regions. In the way of illustration we shall give a few details concerning the work of LITTLEWOOD.

7.19. We shall refer in the sequel to the inequality of HÖLDER: if f and g are non-negative functions, and if $p > 0$, $q > 0$ are exponents such that $(1/p) + (1/q) = 1$, then $\int fg \leq (\int f^p)^{1/p} (\int g^q)^{1/q}$, whenever the integrals involved exist in the LEBESGUE sense (for a particularly elegant proof, see RIESZ [6]).

7.20. Suppose that u is subharmonic for $x^2 + y^2 < 1$ and that $L(|u|^p; 0, 0; \varrho) < G^p$ for $\varrho < 1$, where G is a constant and $p > 1$. If u were *harmonic*, then these assumptions would imply the existence of a function $w(\Theta)$ such that $u(\varrho \cos\Theta, \varrho \sin\Theta) \to w(\Theta)$ for $\varrho \to 1$ and almost every Θ, and

$$\int_0^{2\pi} |u(\varrho\cos\Theta, \varrho\sin\Theta) - w(\Theta)|^p\, d\Theta \to 0 \quad \text{for} \quad \varrho \to 1$$

(Riesz [2]). In the case of a general subharmonic function Littlewood obtained the following results.

7. 21. Under the assumptions of 7. 20 there exists a function $w(\Theta)$
such that

$$\int_0^{2\pi} |u(\varrho\cos\Theta, \varrho\sin\Theta) - w(\Theta)|^q\, d\Theta \to 0, \quad \varrho \to 1,$$

for every exponent $0 < q < p$ (Littlewood [2]). Proof. On account
of the inequality of Hölder it is sufficient to consider the case
$1 < q < p$. Denote by $\bar{h}_r$ the B. H. M. of u for $x^2 + y^2 < r^2$. It
follows by the inequality of Hölder from the formula of Poisson
for $\bar{h}_r$ (cf. 5. 4) that $\bar{h}_r$ satisfies an inequality of the same form as u.
It follows that for $r \to 1$ the sequence $\bar{h}_r$ cannot converge to $+\infty$
everywhere. Hence (see 5. 7 and 1. 4) there exists for u a L. H. M.
h^* in $x^2 + y^2 < 1$ which satisfies an inequality of the same form as u.
By the theorem of F. Riesz quoted in 7. 20 we have therefore a function $w(\Theta)$ such that

$$\int_0^{2\pi} |h^*(\varrho\cos\Theta, \varrho\sin\Theta) - w(\Theta)|^p\, d\Theta \to 0$$

for $\varrho \to 1$. As $u \leqq \bar{h}_r \nearrow h^*$ for $r \nearrow 1$, it follows from the preceding
facts, by repeated application of the inequality of Hölder, that $w(\Theta)$
satisfies the theorem.

7. 22. If u is subharmonic in $x^2 + y^2 < 1$ and $L(|u|; 0, 0; \varrho) < M$
for $\varrho < 1$ (M a finite constant), then $\lim u(\varrho\cos\Theta, \varrho\sin\Theta)$, $\varrho \to 1$,
exists and is finite for almost every Θ (Littlewood [3]). This theorem
is related to the theorems in Chapter VI as follows. Establish first
the existence of the L. H. M. h^* of u in $x^2 + y^2 < 1$ as in 5. 21. Observe next that the assumption concerning u implies that $L(\overset{+}{u}; 0, 0; \varrho)$
is also bounded for $\varrho < 0$. Hence, for the same reasons as in the
case of u, there exists a L. H. M. H^* for $\overset{+}{u}$ in $x^2 + y^2 < 1$, and we
have $H^* \geqq \overset{+}{u} \geqq 0$, $H^* \geqq h^*$ in $x^2 + y^2 < 1$. By 6. 20 we have for u
the representation

$$u(P) = -\int_{x^2+y^2<1} \mathfrak{G}(P, Q)\, d\overset{\circ}{\mu}(e_Q) + h^*(P) = v(P) + h^*(P)$$

for $x^2 + y^2 < 1$, where $\mathfrak{G}$ is Green's function for the unit circle. We
can write $h^* = H^* - (H^* - h^*)$. Thus h^* appears as the difference
of two *non-negative* harmonic functions, and hence h^* has a definite

finite radial limit along almost every radius (see, for references covering also the case of more than two variables, GARRETT [1]). Thus the problem is reduced to the discussion of $v(P)$. LITTLEWOOD shows that $v(P)$ has a radial limit equal to zero for almost every radius. The proof depends upon a number of inequalities concerning GREEN'S function of the unit circle.

7. 23. LITTLEWOOD constructed explicit examples which show that 1) in the theorem of 7. 21 the condition $q < p$ cannot be replaced by $q \leq p$, 2) the theorem of 7. 21 is not valid for $q = p = 1$, and 3) for $0 < p < 1$ there does not exist, generally, a radial limit $w(\Theta)$, either in the sense of convergence almost everywhere or in the sense of convergence in the mean with respect to some exponent (LITTLEWOOD [4]).

7. 24. Theorems on harmonic functions may involve *pairs of conjugate harmonic functions*, that is analytic functions of a complex variable. It is not clear a priori that such theorems can be extended to subharmonic functions of two or more variables. Questions of this type were discussed in considerable detail by PRIVALOFF, who generalized a number of theorems concerned with analytic functions of a complex variable (PRIVALOFF [3, 4]). In the way of illustration, we quote two of his theorems for the case of three independent variables. *Theorem.* Let u be subharmonic in a domain G in three-dimensional EUCLIDean space. Suppose that the boundary B of G can be divided into two parts B_1, B_2 in such a way that $\overline{\lim} u(P) \leq M_k$ if P approaches any point of B_k, $k = 1, 2$. Let $G' + B'$ be a region comprised in G. Then there exist two constants s and t, $0 < s < 1$, $0 < t < 1$, depending only upon G and $G' + B'$, such that in $G' + B'$ we have $u \leq t M_1 + (1 - t) M_2$ if $M_1 \leq M_2$ and $u \leq s M_1 + (1 - s) M_2$ if $M_1 \geq M_2$. *Theorem.* Let u be subharmonic in a domain G in three-dimensional EUCLIDean space. Suppose that there exists a finite constant M such that $\overline{\lim} u(P) \leq M$ if P approaches any point on the boundary of G, with the possible exception of a denumerable set of boundary points Q_n, $n = 1, 2, \ldots$ At these exceptional points it is known that $u(P) - \sigma/(PQ_n) \to -\infty$ for *every* $\sigma > 0$ if P approaches Q_n. Then $u \leq M$ in G (this wording, due to SAKS, is somewhat more general than the original wording of PRIVALOFF). Let us sketch a simple proof (due to SAKS) of the second theorem. Consider in G the

$$\text{function } u_\varepsilon(P) = u(P) - \sum_{n=1}^{\infty} \varepsilon/(2^n PQ_n), \quad \varepsilon > 0.$$ Clearly, the infinite

series converges in G, the convergence being uniform in every region $G' + B' \subset G$. Hence, by 3. 3 and 2. 15, u_ε is subharmonic in G. Clearly, $\overline{\lim} u_\varepsilon(P) \leq M$ if P approaches *any* boundary point of G. By 1. 15 we have therefore $u_\varepsilon(P) \leq M$ in G. For P fixed and $\varepsilon \to 0$ it follows that $u \leq M$ in G.

7.·25. To illustrate results of a different type, we quote the following theorem. Let Γ be a circle and C a convex curve strictly interior to Γ. Suppose that u is positive, continuous and subharmonic in and on Γ. Then

$$\int_C u^\lambda\, ds \leqq 4 \int_\Gamma u^\lambda\, ds \ \text{ for } \lambda \geqq 2, \quad \text{and} \quad \int_C u^\lambda\, ds \leqq \frac{A}{\lambda-1} \int_\Gamma u^\lambda\, ds \ \text{ for } 1 < \lambda < 2,$$

where A is an absolute constant (FRAZER [1]; this is a generalization of previous results of GABRIEL [1, 2], who generalized earlier results of FEJÉR-RIESZ [1]. See also RIESZ [8]). Proof. We can clearly assume that Γ is the unit circle $x^2 + y^2 = 1$. Suppose first that $\lambda = 2$. Denote by $\bar{h}$ the B. H. M. of u in Γ. In Γ we have an analytic function $f(z)$ whose real part is equal to $\bar{h}$, say $f(z) = \bar{h} + ih$, $z = x + iy$. We can suppose that $h(O) = \bar{h}(O)$. If Γ_r is a concentric circle with radius r, such that r is slightly less than 1, then we have by a theorem of GABRIEL [1]

$$\int_C |f|^2\, ds \leqq 2 \int_{\Gamma_r} |f|^2\, ds.$$

We have $\Re f(0)^2 = 0$, since $h(O) = \bar{h}(O)$. Using Γ_r to express $f(0)^2$ by the formula of CAUCHY, we obtain

$$\int_{\Gamma_r} \bar{h}^2\, ds = \int_{\Gamma_r} h^2\, ds.$$

We can write now

$$\int_C u^2\, ds \leqq \int_C \bar{h}^2\, ds \leqq \int_C |f|^2\, ds \leqq 2\int_{\Gamma_r} |f|^2\, ds = 4\int_{\Gamma_r} \bar{h}^2\, ds \underset{r\to 1}{\longrightarrow} 4\int_\Gamma \bar{h}^2\, ds = 4\int_\Gamma u^2\, ds,$$

and the theorem is proved for the special case $\lambda = 2$. If $\lambda > 2$, then apply the preceding result to $u^{\lambda/2}$ which is subharmonic by 3.23. The case $1 < \lambda < 2$ is discussed in a similar fashion. For further theorems of this type see FRAZER [2, 3, 4].

7. 26. Clearly, the method used in 7.25 does not apply in the case of more than two variables. To illustrate a somewhat different situation, we consider a result obtained by SAKS [1] as a corollary to more general theorems. Denote by G a *simply connected* domain in the interior K of the unit circle $x^2 + y^2 = 1$. Let u be subharmonic in G and suppose that $u(P) \to -\infty$ if u approaches any boundary point of G which is *interior* to K. Then $G \equiv K$. This theorem is closely related to recent results of EVANS [5]. The method of EVANS suggests the following proof. Define a function $\bar{u}$ as follows: $\bar{u} = u$ in G and $\bar{u} = -\infty$ in $K - G$. Then u is subharmonic in K (see 1.1 and 2.3). Hence the set $K - G$ cannot have *interior* points (see 1.8). If $K \not\equiv G$, then we can assume that the centre of K is not in G. Apply now the transformation $w = \sqrt{z}$, $z = x + iy$ (cf. RADÓ [1], lemma on p. 2). As G is *simply connected*, we have a single-valued analytic branch of

$\sqrt{z}$ in G, and the transformation leads to a domain G' and a subharmonic function u', such that the assumptions of the theorem are satisfied by G' and u', and such that $K - G'$ *does* have interior points. This contradicts a preceding conclusion, and the theorem is proved.

Saks goes on to show that the preceding theorem is *not* valid in three-dimensional space. Example: consider

$$u(x, y, z) = -\int_0^1 \frac{ds}{r}, \quad r = [(x - s)^2 + y^2 + z^2]^{1/2},$$

in the domain G consisting of all points (x, y, z) in $x^2 + y^2 + z^2 < 1$, except the points $0 \leqq x < 1$, $y = 0$, $z = 0$. We have here one of the many instances where the existence of the transformations $w = \sqrt[n]{z}$ leads to theorems in the plane which cannot be extended to spaces of higher dimensions.

References.

The numbers in round brackets denote chapters and paragraphs.

BECKENBACH, E. F.: [1] On subharmonic functions, Duke Math. J., vol. 1, 1935, pp. 481—483 (*2.13*, *4.5*).

BECKENBACH, E. F., and RADÓ, T.: [1] Subharmonic functions and minimal surfaces, Trans. Amer. Math. Soc., vol. 35, 1933, pp. 648—661 (*4.4*).

[2] Subharmonic functions and surfaces of negative curvature, Trans. Amer. Math. Soc., vol. 35, 1933, pp. 662—664 (*3.25*, *3.26*, *4.5*).

BLASCHKE, W.: [1] Ein Mittelwertsatz und eine kennzeichnende Eigenschaft des logarithmischen Potentials, Ber. Verh. Sächs. Akad. Wiss., Leipziger, vol. 68, 1916, pp. 3—7 (*3.7*).

BOREL, E.: [1] Leçons sur les fonctions de variables réelles, Paris, 1905 (*4.6*).

BRELOT, M.: [1] Étude des fonctions sousharmoniques au voisinage d'un point, Actualités scientifiques et industrielles, vol. 139, 1934 (*1.14*, *3.13*, *4.1*, *4.34*, *5.8*, *7.1*, *7.2*, *7.5*, *7.10*, *7.12*, *7.14—7.17*).

CARLEMAN, T.: [1] Zur Theorie der Minimalflächen, Math. Z., vol. 9, 1921, pp. 154 to 160 (*3.26*).

DANIELL, P. J.: [1] A general form of integral, Annals of Math., vol. 19, 1917, pp. 279—294 (*4.23*).

EVANS, G. C.: [1] Fundamental points of potential theory, Rice Institute Pamphlets, vol. 7, 1920, pp. 252—329 (*5.16*).

[2] Discontinuous boundary value problems of the first kind for Poisson's equation, Amer. J. Math., vol. 51, 1929, pp. 1—18 (*7.18*).

[3] Complements of potential theory, part II, Amer. J. Math., vol. 55, 1933, pp. 29—49 (*7.18*).

[4] On potentials of positive mass, part I and II, Trans. Amer. Math. Soc., vol. 37, 1935, pp. 226—253 and vol. 38, 1935, pp. 201—236 (*1.2*,. *4.6*, *5.22*, *5.34*, *6.3*, *6.21*, *6.22*).

[5] Potentials and positively infinite singularities of harmonic functions, Mh. Math. Physik, vol. 43, 1936, pp. 419—424 (*4.34*, *7.26*).

FEJÉR, L., and RIESZ, F.: [1] Über einige funktionentheoretische Ungleichungen, Math. Z., vol. 11, 1921, pp. 305—314 (*7.25*).

FRAZER, H.: [1] Some inequalities concerning the integrals of positive subharmonic functions along curves of certain types, J. London Math. Soc., vol. 6, 1931, pp. 113—117 (*7.25*).

[2] An inequality concerning subharmonic functions in an infinite strip, J. London Math. Soc., vol. 7, 1932, pp. 214—218 (*7.25*).

[3] Further inequalities concerning subharmonic functions, J. London Math. Soc., vol. 7, pp. 284—290 (*7.25*).

[4] On the moduli of regular functions, Proc. London Math. Soc., vol. 36, 1934, pp. 532—546 (*7.25*).

FROSTMAN, O.: [1] Potentiel d'équilibre et capacité des ensembles, avec quelques applications à la théorie des fonctions, Thesis, Lund, 1935 (*6.21*).

GABRIEL, R. M.: [1] Some results concerning the integrals of moduli of regular functions along curves of certain types, Proc. London Math. Soc., vol. 28, 1926, pp. 121—127 (*7.25*).

[2] An inequality concerning the integrals of positive subharmonic functions along certain circles, J. London Math. Soc., vol. 5, 1930, pp. 129—131 (*7.25*).

GARRETT, G. A.: [1] Necessary and sufficient conditions for potentials of single and double layers, Amer. J. Math., vol. 58, 1936, pp. 95—129 (*7.22*).

HAHN, H.: [1] Theorie der reellen Funktionen, Berlin 1921, vol. 1 (*1.3*).

HAUSDORFF, F. [1]: Grundzüge der Mengenlehre, 1. Aufl. Leipzig 1914 (*4.6*).

HARTOGS, F.: [1] Zur Theorie der analytischen Funktionen mehrerer unabhängiger Veränderlichen, Math. Ann., vol. 62, 1906, pp. 1—88 (*3.37*).

KELLOGG, O. D.: [1] Foundations of potential theory, Berlin 1929 (*1.3*, *1.6*, *1.12*, *1.14*, *4.4*, *4.25*, *5.5*, *5.6*, *5.10*, *5.11*, *5.16*, *5.17*, *5.23*, *5.24*, *6.1*, *6.21*).

KOZAKIEWICZ, W.: [1] Un théorème sur les opérations et son application à la théorie des laplaciens généralisés, C. R. Soc. Sci. Varsovie, vol. 26, 1933, pp. 1—7 (*3.7*).

KURATÓWSKI, C.: [1] Une méthode d'élimination des nombres transfinis des raisonnements mathématiques, Fundam. Math., vol. 3, 1922, pp. 76—108 (*4.6*).

LITTLEWOOD, J. E.: [1] On the definition of a subharmonic function, J. London Math. Soc., vol. 2, 1927, pp. 189—192 (*2.3*).

[2] On functions subharmonic in a circle, part I, J. London Math. Soc., vol. 2, 1927, pp. 192—196 (*7.18*, *7.21*).

[3] On functions subharmonic in a circle, part II, Proc. London Math. Soc., vol. 28, 1928, pp. 383—393 (*7.18*, *7.22*).

[4] On functions subharmonic in a circle, part III, Proc. London Math. Soc., vol. 32, 1931, pp. 222—234 (*7.18*, *7.23*).

MALCHAIR, H.: [1] Sur les fonctions sousharmoniques, Mém. Soc. Roy. Sci. Liège, vol. 20, 1935, pp. 1—15 (*3.5*, *3.22*).

[2] Sur les fonctions sousharmoniques, C. R. Soc. Sci. Varsovie, vol. 28, 1936, pp. 71—76.

MONTEL, P.: [1] Sur les suites infinies des fonctions, Ann. École Norm., series 3, vol. 24, 1907, pp. 233—344.

[2] Sur les fonctions convexes et les fonctions sousharmoniques, J. Math. pures appl., series 9, vol. 7, 1928, pp. 29—60 (*2.8*, *2.11*, *2.17*, *3.5*, *3.12*, *3.13*, *3.17*, *3.18*, *3.20*, *3.22*, *3.37*, *4.3*).

[3] Sur les fonctions doublement convexes et les fonctions doublement sousharmoniques, Prakt. Akad. Athenon, vol. 6, 1931, pp. 374—385 (*3.22*).

MORREY, C. B.: [1] A class of representations of manifolds, part I, Amer. J. Math., vol. 55, 1933, pp. 683—707 (*3.32*).

PÓLYA, G., and SZEGÖ, G.: [1] Aufgaben und Lehrsätze aus der Analysis, vol. I, Berlin 1925 (*3.13*, *4.22*).

PRIVALOFF, I.: [1] Sur un problème limite des fonctions sousharmoniques, Rec. math. Soc. math. Moscou, vol. 41, 1934, pp. 3—9 (*7.18*).

[2] Sur un problème limite de la théorie des fonctions, Bull. Acad. Sci. URSS, series 7, number 2, 1935, pp. 301—304 (*7.18*).

[3] Sur certaines questions de la théorie des fonctions subharmoniques et des fonctions analytiques, Rec. math. Soc. math. Moscou, vol. 41, 1935, pp. 527—550 (*4.3*, *7.24*).

[4] Sur la théorie générale des fonctions harmoniques et subharmoniques, Rec. math. Soc. math. Moscou, N. s. [1], 1936, pp. 103—120 (*2.14*, *4.3*, *7.24*).

RADÓ, T.: [1] Über eine nicht fortsetzbare Riemannsche Mannigfaltigkeit, Math. Z., vol. 20, 1925, pp. 1—6 (*7.26*).

[2] Remarque sur les fonctions sousharmoniques, C. R. Acad. Sci. Paris, vol. 186, 1928, pp. 346—348 (*3.12*).

[3] On convex functions, Trans. Amer. Math. Soc., vol. 37, 1935, pp. 266—285 (*3.27*).

[4] On the harmonic majorants of subharmonic functions, Bull. Amer. Math. Soc. vol. 42, 1936, p. 813 (*5.28*, *5.33*, *5.35*).

See also BECKENBACH-RADÓ.

RADON, J.: [1] Theorie und Anwendung der absolut additiven Mengenfunktionen, Akad. Wiss. Wien, S.-B. Math. Nat. Kl., vol. 122, 1913, pp. 1295—1438 $(4.6—4.8,\ 4.11,\ 4.12,\ 4.15,\ 6.4)$.

RIESZ, F.: [1] Sur les valeurs moyennes du module des fonctions harmoniques et des fonctions analytiques, Acta Litt. Sci. Szeged, vol. 1, 1922, pp. 3—8 $(2.16,\ 3.20,\ 4.3)$.

[2] Über die Randwerte einer analytischen Funktion, Math. Z., vol. 19, 1922, pp. 87—95 (7.20).

[3] Sur une inégalité de M. Littlewood dans la théorie des fonctions, Proc. London Math. Soc., vol. 23, 1924 (4.3).

[4] Über subharmonische Funktionen und ihre Rolle in der Funktionentheorie und in der Potentialtheorie, Acta Litt. Sci. Szeged, vol. 2, 1925, pp. 87—100 $(3.37,\ 4.3,\ 4.6)$.

[5] Sur les fonctions subharmoniques et leur rapport à la théorie du potentiel, part I and II, Acta Math., vol. 48, 1926, pp. 329—343, and vol. 54, 1930, pp. 321—360 $(1.1,\ 1.6,\ 1.7,\ 1.9—1.13,\ 1.15,\ 2.4,\ 2.7,\ 2.9,\ 2.10,\ 2.16,\ 2.19,\ 2.24,\ 3.2—3.4,\ 3.6,\ 3.10,\ 4.3,\ 4.6,\ 4.18,\ 4.23,\ 4.26,\ 4.31,\ 4.34,\ 5.3—5.8,\ 5.10—5.14,\ 5.18,\ 5.20—5.22,\ 6.3,\ 6.4,\ 6.10,\ 6.15,\ 6.18,\ 6.19,\ 7.5)$.

[6] Su alcune disuguaglianze, Boll. Un. mat. Ital., vol. 7, 1928, pp. 1—3 (7.20).

[7] Sur les valeurs moyennes des fonctions, J. London Math. Soc., vol. 5, 1930, pp. 120—121 (2.12).

[8] Eine Ungleichung für harmonische Funktionen, Mh. Math. Physik, vol. 43, 1936, pp. 401—406 (7.25).

See also FEJÉR-RIESZ.

SAKS, S.: [1] Sur une inégalité de la théorie des fonctions, Acta Litt. Sci. Szeged, vol. 4, 1924, pp. 51—55 (7.26).

[2] Sur un théorème de M. Montel, C. R. Acad. Sci. Paris, vol. 187, 1928, pp. 276—277 (3.17).

[3] On subharmonic functions, Acta Litt. Sci. Szeged, vol. 5, 1932, pp. 187 to 193 $(3.7,\ 3.15,\ 3.16)$.

[4] Note on defining properties of harmonic functions, Bull. Amer. Math. Soc., vol. 38, 1932, pp. 380—382.

[5] Théorie de l'intégrale, Warszawa 1933 $(1.4,\ 3.7)$.

SZPILRAJN, E.: [1] Remarques sur les fonctions sousharmoniques, Ann. of Math., vol. 34, 1933, pp. 588—594 $(3.7,\ 3.28,\ 3.33,\ 3.34,\ 3.35,\ 3.36)$.

THOMPSON, J. M.: [1] Distribution of mass for averages of Newtonian potential functions, Bull. Amer. Math. Soc., vol. 41, 1935, pp. 744—752 (6.23).

VALIRON, G.: [1] Remarques sur certaines fonctions convexes, Proc. Phys.-Math. Soc. Jap., series 3, vol. 13, 1931, pp. 19—38 (3.22).

WIENER, N.: [1] Laplacians and continuous linear functionals, Acta Litt. Sci. Szeged, vol. 3, 1926, pp. 7—16 (6.20).